Ergonomics and Human Factors Engineering

Joel M. Haight, Editor

Copyright © 2012 by the American Society of Safety Professionals
All rights reserved.

Copyright, Waiver of First Sale Doctrine
All rights reserved. No part of this work may be reproduced or transmitted in any form or by any means, electronic or mechanical for commercial purposes, without the permission in writing from the Publisher. All requests for permission to reproduce material from this work should be directed to: The American Society of Safety Professionals, ATTN: Manager of Technical Publications, 520 N. Northwest Highway, Park Ridge, IL 60068.

Disclaimer
While the publisher and authors have used their best efforts in preparing this book, they make no representations or warranties with respect to the accuracy or completeness of the contents, and specifically disclaim any implied warranties of fitness for a particular purpose. The information herein is provided with the understanding that the authors are not hereby engaged in rendering professional or legal services. The mention of any specific products herein does not constitute an endorsement or recommendation by the American Society of Safety Professionals, and was done solely at the discretion of the author. Note that any excerpt from a National Fire Protection Association Standard reprinted herein is not the complete and official position of the NFPA on the referenced subject(s), which is represented only by the cited Standard in its entirety.

Library of Congress Cataloging-in-Publication Data

Ergonomics and human factors engineering / Joel M. Haight, editor.
    pages cm
 Includes bibliographical references and index.
 ISBN 978-1-885581-73-0 (alk. paper)
1. Human Engineering. I. Haight, Joel M., editor of compilation. II. American Society of Safety Engineers.
TA166.E7165 2012
620.8'2--dc23
                     2012043232

Notice of Name Change
The American Society of Safety Engineers (ASSE) is now the American Society of Safety Professionals (ASSP).

Managing Editor: Michael F. Burditt, ASSP
Editor: Jeri Ann Stucka, ASSP
Text design and composition: Cathy Lombardi
Cover design: Image Graphics

Printed in the United States of America

24 23 22 21 19 18   4 5 6 7 8 9 10 11

Ergonomics and Human Factors Engineering

Contents

Foreword	iv
About the Editor and Authors	vi
Chapter 1: Regulatory Issues Carol Stuart-Buttle	1
Chapter 2: Principles of Ergonomics Magdy Akladios	21
Chapter 3: Work Physiology Carter J. Kerk and Adam K. Piper	47
Chapter 4: Principles of Human Factors Steven F. Wiker	69
Chapter 5: Benchmarking and Performance Criteria William Coffey	109
Chapter 6: Best Practices Farhad Booeshaghi	127
Index	149

Foreword

THE TERM ERGONOMICS STEMS from the Greek work *ergos*, meaning work, and *nomos*, meaning laws, which translates into *laws of work*. Ergonomics aims to fit the task to the human rather than that other way around. Although ergonomics has historically been used synonymously with human factors, they are becoming two distinct sciences. Ergonomics studies the physiological effects of work activities on people; human factors deals with the interaction between human beings and their work environment, which includes machines and other equipment. Applications of both can often reduce and possibly eliminate potential injuries, accommodate and enhance human performance, and provide an environment where humans and machines work seamlessly and in harmony.

The chapter, "Principles of Ergonomics," gives an overview of the history and basic tenets of ergonomics. The chapter also distinguishes between cumulative trauma disorders (CTDs) and musculoskeletal disorders (MSDs) and distinguishes the various types of ergonomic injuries. Biomechanics is also discussed, as is the lifting formula designated by NIOSH, as well as other tools for analyzing and quantifying exposures to ergonomic stressors. Steven Wiker presents the definition and scope of human factors engineering (HFE) that demonstrates the usefulness of HFE methods in preventing or reducing safety problems through improved design. HFEs work on a variety of design issues and problems that focus on human-machine-task-environment system safety. A general process for avoiding perceptual, cognitive, and motor-related design flaws is also described. Basic computations and models for use in HFE design and review are examined, and the necessity for HFE guidance in all stages of design is discussed. Lastly, sources of information for HFE design guidance are listed for further research.

Another topic which is important to safety engineers is work physiology, which is the study of physiological information about humans and how to apply that information in the evaluation and design of work. In "Work Physiology," Carter Kerk and Adam R. Piper discuss the principles of work physiology and anthropometry, applying those design principles in work design using appropriate data and allowances. Work physiology and anthropometry help the safety professional to minimize occupational injuries, while providing a safer workplace and improving productivity. Fundamental knowledge of body systems, such as the skeletal, skeletal muscular, neuromuscular, respiratory, circulatory and metabolic, are necessary in order to apply work physiology to the evaluation and design of work, which is also examined in this chapter. Also covered are aerobic and anaerobic processes, which are necessary in order to produce energy for work and prevent muscular fatigue. Evaluation of cardiovascular capacity to determine safe and effective job placement is also discussed. The obstacles to safe work design such as physical challenges, age and gender differences and work schedules (night or shift work) are also discussed, giving suggestions of ways to accommodate these differences while still increasing productivity and improving quality.

The question for most safety managers is how you assess human performance at work, which is discussed in this book in terms of benchmarking and performance criteria in Chapter 5.

The author, William Coffey, begins by examining different performance measurement systems, as well as specific metrics. Essentials for setting up an internal benchmarking system are also discussed as well as some potential pitfalls. The financial impact for companies who fail to implement ergonomics programs can be judged by examining their worker' compensation costs, with certain caveats. Both lagging and leading metric indicators are examined, including job/task analysis. Fundamental issues such as how to define a task are discussed. The use of pre-employment screening and functional capacity evaluation (FCE) can be viewed in terms of a metric function as well, with some modification.

Best practices to reduce or eliminate ergonomic hazards and repetitive strains (EHRS), discussed in the last chapter, focuses on the human-machine interface using a human machine task allocation (HMTA). HMTA examines the characteristics of a tool and the human thought processes and decision-making involved in a using a tool. The OSHA guidelines for selecting tools discussed should be carefully considered when selecting hand tools. The OSHA guidelines will reduce worker strain and increase productivity.

Using a tool repeatedly to complete tasks involving limited motion frequently results in injuries termed "repetitive motion injuries." These are the single largest type of occupational health hazard in the United States and reducing or eliminating a task proven to result in such an injury frequently requires a complete examination of a manufacturing process and its redesign. Vibration and noise from the tool can also affect worker health as well as productivity. The reasons both are of major concern and what can be done to reduce or eliminate these hazards should be well understood by safety and health managers. Any workplace environmental factor, which impacts worker performance, including poor lighting, exposure to excessive heat or cold, and workstation design all fall within the realm of human factors engineering. Understanding and addressing these factors, and others, which affect worker health and productivity, require a basic understanding of ergonomics and human factors engineering.

ABOUT THE EDITOR

In 2009, Joel M. Haight, Ph.D., P.E., was named Branch Chief of the Human Factors Branch at the Centers for Disease Control and Prevention (CDC)—National Institute of Occupational Safety and Health (NIOSH) at their Pittsburgh Office of Mine Safety and Health Research. He continues in this role. In 2000, Dr. Haight received a faculty appointment and served as Associate Professor of Energy and Mineral Engineering at the Pennsylvania State University. He also worked as a manager and engineer for the Chevron Corporation domestically and internationally for eighteen years prior to joining the faculty at Penn State. Hehasa Ph.D. (1999) and Master's degree (1994) in Industrial and System Engineering, both from Auburn University. Dr. Haight does human error, process optimization, and intervention effectiveness research. He is a professional member of the American Society of Safety Engineers (where he serves as Federal Liaison to the Board of Trustees and the ASSE Foundation Research Committee Chair), the American Industrial Hygiene Association (AIHA), and the Human Factors and Ergonomics Society (HFES). He has published more than 30 peer-reviewed scientific journal articles and book chapters and is a co-author and the editor-in-chief of ASSE's *The Safety Professionals Handbook* and the John Wiley and Sons, *Handbook of Loss Prevention Engineering*.

ABOUT THE AUTHORS

Magdy Akladios, P.E., CSHM, CSP, is an Associate Professor at the University of Houston–Clear Lake (UHCL) from 2011–present, before which he was an Assistant Professor at UHCL from 2005 to 2011. Prior to coming to UHCL, he was an Assistant, then Associate Professor at West Virginia University from 1996–2005. Dr. Akladios' education includes a Ph.D. in Industrial Engineering, a Master's degree in Industrial Engineering, a Master's degree in Occupational Safety & Health (Industrial Hygiene), an M.B.A, and a B.S. in Mechanical Engineering. In addition, he is a Board Certified Safety Professional (CSP), a Professional Engineer (PE), a Certified Professional Ergonomist (CPE), a Certified Safety & Health Manager (CSHM), and a Certified Member of the Egyptian Syndicate for Mechanical Engineers, a member of the Human Factors & Ergonomics Society (HFES), and a Senior Member of the Institute of Industrial Engineers (IIE). He is also a member of the Industrial Engineers' Honorary Society (Alpha-Pi-Mu) and an ABET Program Evaluator (PEV).

Farhad Booeshaghi, Ph.D., P.E., is Adjunct Professor at the Florida State University–Florida A&M University College of Engineering and a managing member and consulting engineer at Global Engineering & Scientific Solutions, LLC.

Craig Arthur Brown, M.S., P.E., CSP, has worked as an EHS professional in the oil and gas industry for 30 years.

William R. Coffey, CPEA, CSP, is President, WRC Safety and Risk Consultants. *Section Coordinators/Authors* xiii

Carter J. Kerk, Ph.D., P.E., CPE, CSP, is a Professor in the Industrial Engineering Department at the South Dakota School of Mines.

Adam K. Piper, Ph.D., CSP, is an Assistant Professor in the Industrial Engineering Department at the South Dakota School of Mines.

Carol Stuart-Buttle, CPE, is Principal of Stuart-Buttle Ergonomics, Philadelphia, Pennsylvania.

Steven F. Wiker, Ph.D., CPE, is the Director, Ergonomics Design Institute. He conducts research and consults in the fields of ergonomics, safety, and industrial engineering.

1

REGULATORY ISSUES

Carol Stuart-Buttle

LEARNING OBJECTIVES

- Understand how politics relates to standards-setting in the United States.

- Become familiar with the most important current standards and guidelines.

- Be aware of the current government policy and initiatives on ergonomics.

- Understand the influence of international standards.

- Recognize the relationship between workers' compensation and ergonomics.

- Know where to find standards.

REGULATIONS HAVE ALWAYS BEEN entwined with politics. Early debates in the Congress of the United States raised fundamental differences about the relative control by federal and state governments over any aspect of residents' lives. Today there is debate on regulation versus no regulation, regardless of whether at the federal or state level. Ergonomics regulation has been at the heart of such contention.

Such regulation has drawn opposing views from groups representing business and those representing workers, including unions. The business community, harboring a philosophical resistance to intervention and oversight by the Occupational Safety and Health Administration (OSHA), opposes regulation. It argues that regulation increases costs (Kirchgaessner 2010, Karr 2002, Mugno 2002, and Greenhouse 2001). However, many companies that are members of associations arguing against regulation have established ergonomic initiatives and maintain them. Supporters of regulation state that workers are getting hurt in some industries when ergonomics is ignored, and the "bad apples" in business need regulation because it pushes them to implement measures that protect workers. Sometimes the message that ergonomics is good for business and, if done right, can be cost effective is lost among arguments by the most vocal groups on either side of the debate. Those in safety and health, as well as the human factors and ergonomics communities, state that ergonomics is good economics, as repeatedly manifested by articles in professional and trade publications (Chapman 2006; McDermott, Lopez, and Weiss 2004; Hogrebe 2004; Dwyer and Lotz 2003; Rodrigues 2001; Maloney 2000; Dolan 1998; and Hendrick 1996).

In addition, the General Accounting Office (GAO) (renamed the Government Accountability Office) has reported positive results from ergonomics programs (GAO 1997).

One troubling aspect of politicizing ergonomics was exposed when, in the heat of fierce rhetoric over a federal standard, the field of ergonomics was portrayed as having no foundation in science. Nevertheless, Congress funded the National Research Council and Institute of Medicine (within the National Academy of Sciences) to conduct a scientific review to determine if work-related musculoskeletal disorders were founded in research. It concluded there is a scientific basis for both injury risk and interventions (NRC 2001).

But the political battles continue. The federal Ergonomics Standard of 2000 was rescinded a few months after its passage with the change of administration from President Clinton to President Bush.

Some of the same business groups involved in the federal debates also took on the Washington State Ergonomics Standard. Lawsuits were attempted on the basis that the regulation was invalid (BNA 2001). When that did not succeed, the opposition brought the standard to popular vote and advocated voting against it, resulting in its defeat.

Similarly, politics has affected voluntary standards. Groups opposing ASC Z-365, *Management of Work-Related Musculoskeletal Disorders*, introduced legal tactics that made the financial burden on the secretariat of the potential standard too great to carry (BNA 2003). The committee was discontinued. A California ergonomics rule has weathered petitions from both business communities opposing the rule and union groups stating the rule is too lenient. The most recent petition from the California Labor Federation has stalled and remains under review by an advisory committee (BNA 2004a).

In 1998, R. D. Fulwiler succinctly stated that regulatory pressure is no longer the driver for safety and health as it once was, and it never should have been a key driver. However, many companies need the regulatory nudge to put mechanisms in place that support both business success and health and safety success (Fulwiler 1998). Since Fulwiler's statement, ergonomics has been less in the foreground until recently; Ful-

wiler's words are once again pertinent. Activities addressing MSDs have increased. OSHA has hired more inspectors to raise the volume of targeted inspections of industry. There is also a proposal to enhance record keeping that includes adding a column to capture MSDs on the OSHA log. Hilda Solis, Secretary of Labor in the Obama administration since 2009, denies that these actions are a "prelude to a broader ergonomics standard" (BNA 2009).

These OSHA activities have reactivated the political parties that debated the federal ergonomics standard. Those who oppose the record-keeping proposal are typically businesses claiming that the proposal is a precursor to passing an ergonomics standard. Those who support the proposed OSHA changes include labor unions, stating that there is a need for better tracking and for addressing musculoskeletal problems through ergonomics (BNA 2009). Membership in labor unions has decreased dramatically since the 1980s, which makes it difficult for unions to have a voice in such issues. Some large safety and health associations, such as the American Society of Safety Engineers (ASSE) and the American Industrial Hygiene Association (AIHA), often support a position similar to that of the unions.

STANDARDS IN THE UNITED STATES
OSHA Standards and Guidelines

The primary mandatory standard in the United States is the Occupational Safety and Health (OSH) Act of 1970 (OSHA 1970). Citations for ergonomic violations are based on the General Duty Clause of the Act [Section 5(a)(1)], which states: "Each employer shall furnish to each of his employees, employment and a place of employment which is free from recognized hazards that are causing or are likely to cause death or serious harm to his employees."

This high-level statement has led to contests over ergonomic citations, especially questioning what is a recognized hazard and whether some of the medical conditions that might arise from poor ergonomic design cause serious harm. There is no other federal standard specific to ergonomics. Despite contests against OSHA, some landmark cases (Pepperidge Farms and

Regulatory Issues

Beverly Enterprises) have been settled by courts that interpreted the general duty clause pertaining to ergonomics, particularly related to risk factors of repetitive-motion disorders and lifting (Abrams 2002, BNA 1997 and 2002). The outcome of these cases has made it possible for OSHA to continue issuing ergonomic citations with fewer contests arising and to reach settlement agreements with companies. From 2002 to 2004 OSHA conducted 1500 ergonomic inspections and issued thirteen ergonomic citations using the General Duty Clause (BNA 2004b). From 2004 to 2010 there has been only one federal citation for ergonomics under the General Duty Clause, and that was in Puerto Rico (OSHA 2010).

Ergonomics Program Standard Repealed

OSHA released an Ergonomics Program rule in November 2000 that was repealed in March 2001. The standard was published in the *Federal Register* on November 14, 2000, Vol. 65, No. 220 (29 CFR 1910.900) and is still available in public records (Federal Register 2000a). The standard was developed over approximately ten years. The early drafts were prescriptive and detailed, while the final released version was more programmatic. Early drafts mandated that companies establish an ergonomics program, whereas the final version provided a quick fix that, if effective, would not require a full ergonomics program. An extensive appendix provided guidance material on how to meet the standard.

The main elements of the standard were:

1. Provide basic information to employees on musculoskeletal disorders (MSDs).
2. Determine the work-relatedness of an employee's reported MSD.
3. Provide prompt medical management of work-related MSDs.
4. Implement a quick fix if two or less cases on the same job, or go to a full program.
5. When more than two cases arise or a quick fix is ineffective, implement a full ergonomics program with the following elements:
   - management leadership
   - employee participation
   - job hazard analysis
   - hazard reduction and control
   - training
   - program evaluation.

Although the standard was rescinded, it might be useful to an industry as a guide for addressing its ergonomic issues. Very similar material is also available on the Washington state government Web site related to their state ergonomics standard, which was also repealed. (See the Washington State Ergonomics Standard later in this chapter.)

OSHA's Four-Pronged Approach to Ergonomics

After the repeal of the ergonomics standard in 2001, the Bush administration charged the Department of Labor (DOL) with addressing ergonomic-related problems, usually termed occupational musculoskeletal disorders (MSDs). The Department of Labor, under Secretary Elaine Chao, conducted public hearings and stakeholder meetings, received comments, and studied alternative approaches to address ergonomic-related problems. A DOL news release on April 5, 2002 announced the DOL plan (OSHA 2002a and 2002b). The four-pronged approach of the DOL plan outlined the steps OSHA would take to address MSDs. The plan is *not* a rule or standard, and the four-pronged approach remains in place in 2010. The four parts are:

a. *Guidelines*. Industry-specific, voluntary guidelines are being developed to provide effective and feasible solutions to ergonomic-related problems. The injury and illness incidence rates of an industry are determining factors for which industries need guidelines. (For more details, see "Guidelines Affecting Ergonomics.")

b. *Enforcement*. OSHA continues to inspect facilities for ergonomic-related problems and issue citations or hazard alert letters based on the General Duty Clause of the OSH Act. In addition, OSHA conducts a Site Specific Targeting (SST) Inspection Program that identifies workplaces with high incident rates (annual rates based on 100 full-time workers). The list is generated annually based on the latest injury data. Identified facilities receive a notice that their incidence rates are high and they could be inspected. There are three groups with different target criteria: (1) manufacturing, (2) nonmanufacturing, and (3) nursing homes and personal care facilities (OSHA 2009a). In 2009, manufacturing facilities with injury or illness rates at or greater than eight, resulting in days away from work, restricted work activity, or job transfer for every 100 full-time workers [known as the days away, restricted, or transferred (DART) rate] were sent letters. Nonmanufacturing sites with DART rates of 3 or more received notices. The primary SST list also includes sites that have a days away from work injury and illness (DAFWII) rate of 6 or higher for manufacturing, and 13 or higher for nonmanufacturing facilities. The average national DART rate in 2007 for private industry was 2.1, while the national average DAFWII rate was 1.2. The SST list stems from OSHA's Data Initiative for the previous year, which surveys approximately 80,000 employers to attain the latest injury and illness data (OSHA 2009a).

c. *Outreach and Assistance*. Many compliance tools and information are provided through the OSHA Web page (www.osha.gov), including "e-tools" on specific topics and industries, case studies, and training and education resources. The administration also has developed cooperative programs that include: Alliances, Consultation, the Safety and Health Achievement Recognition Program (SHARP), Strategic Partnerships, and the Voluntary Protection Program (VPP). All of these programs existed before the four-pronged initiative, except for Alliances (formed in 2002), but the DOL plan has promoted cooperation. VPP and SHARP guide and encourage high standards of safety and health processes that include ergonomics. Alliances and Strategic Partnerships forge bonds between OSHA and many industries and associations. Alliances are formal agreements with groups and companies to cooperate on developing resources to address workplace safety and health. Currently, there are 70 national OSHA office alliances, but regional office alliances can also be formed. Of the national alliances, 26 are categorized as having an ergonomic emphasis. OSHA Strategic Partnerships (OSPs) have existed since 1998 but have grown under the four-pronged plan. OSPs are entered into with companies that want help in addressing specific safety and health issues. At the end of March 2005, OSHA had formed 373 partnerships, 228 of which remain open. Eighteen OSPs specifically emphasized ergonomics. However, ergonomics is frequently an important part of all these cooperative programs (OSHA 2005a).

d. *National Advisory Committee on Ergonomics (NACE)*. This committee was formed at its initial meeting in January 2003 and was chartered until November 2004. The committee was charged with the following:
- provided information related to various industry or task-specific guidelines
- identified gaps in the existing research base related to applying ergonomic principles to the workplace
- identified current and projected research needs and efforts
- determined methods of providing outreach and assistance that will communicate the value of ergonomics to employers and employees
- provided ways to increase communication among stakeholders on the issue of ergonomics.

OSHA's Six-Year Strategic Plan

Secretary Hilda Solis announced during a Web chat that OSHA is developing a six-year strategic plan that is expected to be in place by September 2010. The overview states that OSHA plans to promote safety and health in work environments by "setting and enforcing workplace safety and health standards; delivering effective enforcement; providing outreach education and compliance assistance; and encouraging continual improvement in workplace safety and health" (Walter 2010).

Guidelines Affecting Ergonomics

There are two guidelines that historically have influenced ergonomics: the Safety and Health Management Guidelines (Federal Register 1989) and the Ergonomics Program Management Guidelines for Meatpacking Plants (OSHA 1990). In addition, under the recent four-pronged initiative, OSHA issued three industry guidelines and another is under development. The following introduces each of these guidelines in more detail.

   a. *Safety and Health Management Guidelines.* These guidelines were issued in 1989 as a result of successful Voluntary Protection Programs (VPP) (Federal Register 2000b). The guidelines provide criteria for organizing a managed safety and health program (Federal Register 1989). Despite the age of these guidelines, they are still referred to, and subsequent guidelines such as the Ergonomics Program Management Guidelines for Meatpacking Plants (OSHA 1990) have been based on them. The SHARP program also refers to these guidelines as the minimum standard from which to develop and maintain a safety and health management system (OSHA 2005b).

   The guidelines are general and have four elements: management commitment and employee involvement, work-site analysis, hazard prevention and control, and safety and health training. The preamble emphasizes that an effective program is more important than a written one but that written guidance enhances communication.

   b. *Ergonomics Program Management Guidelines for Meatpacking Plants.* OSHA issued these guidelines in 1990 in response to the high incidence of cumulative trauma disorders in the red meatpacking industry (OSHA 1990). The guidelines emphasize management commitment and employee involvement and cover four main program elements: work-site analysis, hazard prevention and control, medical management, and training and education. Medical management was introduced as a program element because of the severity of the musculoskeletal disorders that were found in the meatpacking industry at the time. The guidelines also offer detail on implementation and examples for red meatpacking. At around this time, the press reported a clamor by industry and some in Congress for guidance to control the "national epidemic" of repetitive-motion injuries, and a draft federal standard was then expected from OSHA fairly soon after the guidelines (BNA 1991, Occupational Hazards 1990). Other industries have often referred to the meatpackers' guidelines, especially since it was several years until there was a drafted federal ergonomics standard.

   c. *Industry Guidelines.* Under the four-pronged OSHA initiative, three guidelines have been released in final versions and a fourth is being developed. NACE made recommendations of industries that might benefit from guidelines, and OSHA is expected to create more, but none has been announced to date. Those that have been developed are anticipated to be useful to industries other than the target audience. The guides are heavily illustrated and emphasize identifying and controlling hazards that are common in the industry. Copies are available at www.osha.gov. The guidelines in order of release are:

   - guidelines for nursing homes
   - guidelines for retail grocery stores
   - guidelines for poultry processing
   - guidelines for shipyards

VPP and SHARP

The Voluntary Protection Program (VPP) and the Safety and Health Achievement Recognition Program (SHARP) are both successful national incentive programs for improving safety and health. To be recognized by either program, a company is required to have injury and illness rates below the national average. The head of OSHA in 2002, John Henshaw, announced plans to increase company participation and stated that VPP work sites have 50 percent fewer workplace injuries than non-VPP work sites (BNA 2004c, Safety and Health 2002). VPP and SHARP are listed as part of the overall cooperative programs that are elements of the outreach and assistance prong of the OSHA ergonomic initiative. To demonstrate excellence in safety and health, a company needs to attain and maintain a low injury rate. To achieve that, many companies must address MSDs through ergonomics.

The VPP was established in 1982 and its requirements center on comprehensive management systems with active employee involvement to prevent or control the safety and health hazards at the work site (Federal Register 2000b). In 2000, these VPP guidelines were reorganized to be consistent with the 1989 Safety and Health Management Guidelines. As a result, there are now four rather than six elements to the VPP. This program can provide a safety and health incentive for a company, as well as support mechanisms and mentoring from other VPP companies, to help reach the high standard. In addition, sites are removed from programmed OSHA inspections during VPP participation. However, OSHA is still obliged to respond to any employee complaint or any fatality or catastrophe. There are three levels of VPP programs: Star, Demonstration, and Merit. Star recognizes workplaces that are self-sufficient in controlling hazards at the work site. The Demonstration level includes work sites with Star-quality safety and health programs but that require demonstration or testing of experimental approaches that are different from Star. The Merit level recognizes good safety and health programs, but additional steps are required to reach the Star level. The 2000 Federal Register notice announced more relaxed application criteria for VPPs to encourage companies to apply for the program and then strive to reach the Merit or Star status, which still requires meeting stringent criteria. Previously, the program posed tougher criteria for entry. In 2004, VPPs were initiated to promote safety and health in the construction industry because that industry was not eligible for participation under the existing VPP (BNA 2004d). In 2009, OSHA made further changes in acceptance criteria to be more inclusive of companies with a mobile workforce (OSHA 2009b). However, the Government Accountability Office (GAO) released a report the same year concluding that OSHA does not have adequate oversight of the program (GAO 2009). VPP funds have been reduced, and OSHA is evaluating the program (ISHN 2009).

The Safety and Health Achievement Recognition Program (SHARP), is a recognition program designed particularly for small employers. As with the VPP, SHARP exempts the facility from inspection during the time that the SHARP certification is valid, except when an employee complaint is filed or a fatality or catastrophe occurs. To participate in SHARP, a company needs to:

- request a consultation visit that involves a complete hazard identification survey
- involve employees in the consultation process
- correct all hazards identified by the consultant
- implement and maintain a safety and health management system that, at a minimum, addresses OSHA's 1989 Safety and Health Program Management Guidelines
- lower its Lost-Workday Injury and Illness rate (LWDII) and Total Recordable Case Rate (TRCR) below the national average
- agree to notify the state Consultation Project Office prior to making any changes in working conditions or introducing new hazards into the workplace.

If the requirements are satisfied, the work site will be recommended for approval and certification (OSHA 2005b).

American National Standards Institute

The American National Standards Institute (ANSI) is a voluntary consensus standards body that issues many standards in a year. The following are just a few of ANSI's main ergonomics standards that exist or are being developed. There are others that relate to safety and ergonomics in the workplace but pertain to more specific domains. Documentation may be purchased through ANSI, but often it is purchased directly from the group responsible for developing the standard in coordination with ANSI. Although voluntary, at times ANSI standards are cited in OSHA regulations.

Recently federal agencies and departments have moved toward adopting private-sector standards rather than military standards. This change of direction was provoked by laws such as the National Technology Transfer and Advancement Act of 1995 that encouraged all federal agencies and departments to use technical standards published by voluntary consensus standards bodies as much as possible (Public Law 2006).

The following are some of the general ergonomics standards that ANSI has released or has under development that are less industry specific.

a. *ANSI/HFES 100-2007, Human Factors Engineering of Computer Workstations.* There are four main sections: Installed Systems, Input Devices, Visual Displays, and Furniture. The standard provides basics on ergonomics of computer workstations, including workstation design specifications for the anthropometric range of 5th to 95th percentile and for four reference postures: reclined, upright, and declined sitting, and standing. Several input devices are covered, and there is a section on how to integrate all the workstation components into an effective system (HFES 2007).

b. *ANSI/AIHA Z10 Occupational Health Safety Management Systems.* The American Industrial Hygiene Association (AIHA) issued ANSI/AIHA Z10 in 2005 (AIHA 2005). This document is applicable to organizations of all sizes and provides a concept and action outline to improve safety and health management systems. All safety and health management standards and guidelines are relevant to ergonomics because ergonomics is successful when conducted within a framework, whether that framework or process is an ergonomic process or integrated into the safety and health process. Other than introductory sections, the standard covers five main topics: management leadership and employee participation, planning, implementation and operation, evaluation and corrective action, and management review. The appendix includes some forms and audit materials as well as an example of a risk-assessment matrix. The standard is a performance standard, not a specification standard, therefore it can be easily integrated into any other existing management systems. Z10 is intentionally compatible with the International Labour Organization's (ILO) Guidelines on Occupational Health and Safety Management Systems (ILO-OSH 2001) (Manuele 2006). AIHA has also published a companion document aligned with ANSI/AIHA Z10-2005 called the "Ergonomics Program Guidance Document." The guidance provided is connected by paragraph number to each section of the standard (Rostykus 2008).

c. ASSE Z790 *Prevention Through Design (PTD): Guidelines for Addressing Occupational Risks in Design and Redesign Processes.* This document is currently an ASSE Technical Report (TR-Z790.001.-2009) and ASSE plans to apply to ANSI to develop the document into a standard (ASSE 2009, ASSE 2010). The technical report delineates the integration of hazard analysis and risk-assessment methods into the early stages of engineering and design of new and redesigned systems, in order to prevent injuries (ASSE 2010).

d. *ANSI/HFES 200 Human Factors Engineering of Software User Interface.* This is a five-part standard issued by the Human Factors and Ergonomics Society (HFES) (HFES 2008). It closely mirrors the International Standards Organization, ISO 9241 standard on visual display terminals, except for original parts on accessibility and interactive voice response. The five sections are: Introduction, Accessibility, Interaction Techniques, Interactive Voice Response, and Visual Presentation—Use of Color.

e. *ASC Z-365 Management of Work-Related Musculoskeletal Disorders—discontinued.* The

ASC Z-365 was formed in 1991. The secretariat, National Safety Council (NSC), issued the most recent working draft that was near completion in October 2000. The draft contained similar elements to that of the repealed federal standard and the OSHA meatpackers' guidelines. The document was programmatic rather than specific in that it did not provide details on how to conduct analyses or implement interventions. In the fall of 2003, the secretariat withdrew from the process citing extraordinary costs involved with this particular standard. After twelve years and with an almost completed standard, the public review completed, and approval pending, the committee was disbanded as no alternative secretariat came forward (BNA 2003).

American Conference of Governmental Industrial Hygienists' TLV®

The American Conference of Governmental Industrial Hygienists (ACGIH) is well known for developing and publishing threshold limit values (TLV) for chemical substances and physical agents. These include TLV for hand-arm vibration, whole-body vibration, and thermal stress. In 2001, a new ergonomics section was published with TLV for hand activity level (HAL) as well as lifting TLV (ACGIH 2001).

The HAL TLV is intended for monotask jobs performed for four hours or more. A rating system is used to assess the average hand activity level and related recovery time. The TLV is determined from a graph of HAL against normalized peak hand force. The TLV depicted by a solid line on the Action Limit graph is based on combinations of force and hand activity associated with a significantly elevated prevalence of musculoskeletal disorders. A lower alternative action limit indicates that some surveillance should be instituted and general controls implemented.

The lifting TLV provides weight limits based on frequency and duration of lift within 30 degrees of the sagittal plane. There are three lifting tables defined by the duration of the lifting and the rate of lift per hour. Each table portrays a grid of three horizontal distances from the low-back and four-height zones of the start of the lift. Weight limits are provided in the matrix for each circumstance. The TLV of weights under the defined conditions of each table are considered to be acceptable values for nearly all workers on a repeated daily basis without developing work-related lower-back or shoulder disorders associated with repetitive-lifting tasks. These tables are based on data that include more recent dynamic biomechanics (Marras and Hamrick 2006).

National Institute of Standards and Technology

The National Institute of Standards and Technology (NIST) helps to develop standards that address measurement accuracy, documentary methods, conformity assessment and accreditation, and information technology standards. A current initiative of NIST is to develop industry usability reporting guidelines that directly affect software ergonomics. NIST can also be a channel through which to link to military standards. Many of the standards are domain specific, such as automotive, chemical processing, construction, healthcare, and manufacturing, and are especially concerned with quality and system efficiency as well as equipment performance.

In 1988 NIST launched The Malcolm Baldrige National Quality Award that recognizes performance excellence and quality achievement by U.S. manufacturers, service companies, educational organizations, and healthcare providers. This award has driven many companies to address safety and health issues, in particular ergonomics, so they can achieve performance excellence and be a contender for the award. In 2010 there were 83 contenders for the award, which is a 20 percent increase in applicants over 2009 (NIST 2010).

Miscellaneous Standards-Setting Groups

There are many other sources of standards relating to ergonomics that may be important to certain domains or specialties. A few of the other organizations that develop standards are the American Society of Mechanical Engineers (ASME), the American Society for Testing and Materials (ASTM), the Institute of Electrical and Electronics Engineers (IEEE), and the Society of Automotive Engineers (SAE).

STATE REGULATIONS

California Ergonomics Standard

This is a mandatory state law, made effective in 1997, which addresses formally diagnosed work-related repetitive-motion injuries that have occurred to more than one employee performing identical work activities (CAL/OSHA, 1997). The employer must implement a program to minimize the repetitive-motion injuries through work-site evaluation, control of the exposures, and training. The standard is nonprescriptive in that it does not state how to perform an evaluation, control the exposures, or conduct training. The brief two-page standard gained some notoriety since it was issued during the same period as early drafts of the federal standard, which were several pages long with extensive supporting material.

Maine's Video Display Terminal Law

Maine has a video display terminal operators law under the health and safety regulations that requires employers to provide an education and training program to all employees who use a computer for four or more hours a day when they are first employed, and annually thereafter (Safetyworks 1989). Although the term *ergonomics* is not used, the training required pertains to ergonomic principles of how to set up a computer workstation for user comfort and health.

Washington State Ergonomics Standard

In the year 2000, Washington state adopted a mandatory ergonomics rule (Labor and Industries 2006a). The rule stated that employers have to look at their jobs to determine if there are specific risk factors that make a job a "caution zone job" as defined by the standard. All caution zone jobs must be analyzed, employees of those jobs are to participate and be educated, and the identified hazards are to be reduced. Although the rule was issued in 2000, there was a two-year grace period for compliance. However, the rule was repealed in November 2003 (effective December 2003) after state residents voted against the standard. Much of the information on the rule remains on the Washington State Labor and Industries Web site as a resource in ergonomics (Washington Department of Labor and Industries 2006b). The information includes resources such as case histories, checklists and calculators for determining the risk for MSDs, publications, and videos.

Michigan's Ergonomics Rule

In June of 2002 a steering committee was formed by two standing commissions of Michigan's OSHA, a General Industry Standards Commission and the Occupational Health Standards Commission. The steering committee was to develop a framework for addressing an ergonomics rule, and an Ergonomics Standard Advisory Committee was formed to address the feasibility of such a standard. Draft #17 (January 14, 2009) of the ergonomics rule, written by the Ergonomics Standard Advisory Committee, has been approved by the state committees and now needs to be approved by the director of Michigan Department of Energy, Labor & Economic Growth. The rule has been kept as minimal as possible, and lays out the minimum requirements for awareness training, and a process for assessing and responding to ergonomic occupational risk factors (Law 2010). In 2004, the Governor of Michigan, Jennifer Granholm, vetoed a bill that banned funding an ergonomics standard (BNA 2005). There have been several attempts to stop the standard, including a bill, S93, proposed to the state congress at the beginning of 2009 (Risk and Insurance 2009). The controversy over a state ergonomics standard is heating up in 2010, since an economic analysis and regulatory impact statement on the proposed standard is nearly completed. There are several steps yet to be taken, including public hearings for the standard to become law. Opponents plan to appeal the proposed law and hold up the ergonomics standard process until gubernatorial elections in November 2010, when they hope for a change of governor who will not support the standard (Laws 2010).

Legislation in Multiple States

Bills on safe patient handling have been passed in nine states (Illinois, Maryland, Minnesota, New Jersey, New York, Ohio, Rhode Island, Texas, and Washington), legislation introduced in another five states (Hawaii, Massachusetts, Michigan, Missouri, and Vermont), and further legislation was passed in New York and Minnesota during 2010 (ANA 2010). The risk of musculoskeletal injuries is high, especially for back injury, where an estimated 52 percent of the nursing workforce has chronic back pain and 12 percent of nurses leave the profession because of back problems (ANA 2010, Anderson 2006). The state bills are similar in that most require establishing a Safe Patient Lifting Program with a committee, annual training, the use of lifting and repositioning aids, and evaluation of the effectiveness of the program (Hudson 2007).

In 2009 two federal bills were proposed to become The Nurse and Health Care Worker Protection Act of 2009 (HR 2381 and SB1788). The bills are similar and include a standard to be issued by OSHA on safe patient handling, requirements for the use of patient lifting equipment, implementation of plans for patient handling, and training of workers (Hudson 2010).

Workers' Compensation in the United States

Injury rates are one of the primary measures of the effectiveness of an ergonomics program. Injuries can indicate a design problem, and reduction in injuries through intervention is one measure of effectiveness. workers' compensation (WC) rates and costs provide a dollar amount related to the injuries in a company. They can be another benchmark along with OSHA log rates. WC costs are also part of return-on-investment calculations. In ergonomic analyses, these WC measures are used along with other indicators, including performance measures, for figuring returns on investment (Oxenburgh and Marlow 2006, Stuart-Buttle 2006a and 2006b).

The WC laws were established in the early 1900s to provide a mechanism for processing work-related injury and illness claims. When an employer accepts the responsibility for an occupational injury, regardless of fault, and the injury is processed through the WC system, the employer is protected from common lawsuits alleging negligence. Although the worker forfeits the right to sue, he or she benefits from economic compensation of a percentage of wages lost while injured as well as medical treatment and rehabilitation. (Refer to Section 7 of this Handbook for an in-depth discussion of workers' compensation.)

There are specific controls within the WC system to prevent abuse (Maygar 1999). They include:

1. Injuries must occur during the course of normal employment.
2. Injured workers must submit to a medical evaluation, although they can refuse medical treatment.
3. Injuries must be reported within a prescribed time period.
4. Medical treatment and benefit limits are established.
5. Maximum medical improvement (MMI) exams are conducted and the extent of residual disability determined on more extensive cases.
6. Informal hearings or conferences are held to resolve disputes and come to mutual agreements with the Workers' Compensation Commission.

Careful recording and close monitoring of each case by the employer helps prevent confusion or disputes. Injury prevention and a prompt, effective, return to work reduces costs through lower WC premiums and reserves that are based on a company's claim history. By practicing injury prevention, a company also saves on indirect costs, such as lost productivity and temporary hiring costs. These cost-reduction strategies are often left to the employer to implement, but the company may need guidance on how to move a worker back to full employment without reinjury. This is best achieved through safety and ergonomic processes (Stuart-Buttle 2006b). (For more on Safety Management, see Section 3 of this handbook.)

The WC system is run at the state level and covers approximately 127 million workers nationally without any congressional role (Treaster 2003). The WC premiums vary by state, as do the related regulations pertaining to safety initiatives. Most of the safety regulations tied in with WC laws were adopted during WC reforms in the 1980s and 1990s in an effort to control occupational injuries and overall WC costs. Just over half the states (52 percent) have at least one type of safety requirement such as, safety committees, safety and health programs, insurance carrier loss-control services, and targeting initiatives (Smitha, Oestenstad, and Brown 2001). Twenty-four states (48 percent) have no safety requirement. Usually the employers are subject to penalties if they fail to comply with mandatory workplace safety requirements. The study by Smitha et al. (2001) of WC safety regulations focused on the four types of safety requirements that were mandatory in a state from 1992 to 1997. From one to three of these types of safety requirements may be present in a single state. Nine states had safety committee requirements, ten had safety and health program requirements, eleven had regulated insurance carrier loss-control services, and eighteen states had targeting initiatives in place, which means there were mandatory safety requirements for employers with above-average injury or WC loss rates. Between 1980 and 1990 claim costs increased an average of 11 percent per year, while between 1991 and 1995 the costs increased only 2 percent. Smitha et al. (2001) attribute this to prevention initiatives under the WC reforms. However, more recent figures reflect the national increase in medical costs that have increased WC costs as well. As reported in 2003, the average cost of WC insurance has risen 50 percent in the last 3 years (Treaster 2003). The costs also vary by state, based on the claims experienced (Geddes Lipold 2003).

Americans with Disabilities Act

The Americans with Disabilities (ADA) Act came into effect in 1990 (EEOC 1990). One part of the act addresses accessibility for the disabled and a second part pertains to employment. Two main points of the act under the employment section are:

- The ADA prohibits disability-based discrimination in hiring practices and working conditions.
- Employers are obligated to make reasonable accommodations for qualified disabled applicants and workers, unless doing so would impose undue hardship on the the employer. The accommodations should allow the employee to perform the essential functions of the job.

The act also has some bearing on ergonomics. Human Resource Departments need to be aware of the essential functions of a job, so many companies have modified their job descriptions to include them. However, in some industries, defining those functions and job demands is not always easy and can entail the assistance of safety and health personnel. Many different personnel in industry, including those working in ergonomics, need job descriptions that often provide information related to their area of responsibility. So those addressing ergonomics should be aware of what descriptions do exist and should be sure that essential functions are included (Stuart-Buttle 2006b).

If a disabled worker needs to be accommodated, the company may make workstation modifications. Often this can provide an opportunity to improve a job for all the workers. A job should be thoroughly assessed so that any changes would provide the necessary accommodation and also make an ergonomic improvement for others. Caution is advised in tailoring a workstation to one person to such an extent that no one else could work in that area if the person is absent. This implies the benefit of including adjustability that helps everyone. Designed appropriately, making the changes should provide a return on the investment rather than generating a cost to accommodate one person (Stuart-Buttle 2006b).

INTERNATIONAL STANDARDS

International Standards Organization

The International Standards Organization (ISO) organizes its standards by general topic area into 40 main groups. There are three types: A and B standards that address principles and processes and C standards that are technically based and mostly of interest to specific types of industries. Technical Committees (TC) also have their own listings of standards. For example, TC 159 is the Ergonomics Technical Committee. Apart from specific ergonomic-related standards, some rules in other categories and TCs may be pertinent to a particular industry, such as the categories of electronics or material-handling equipment. Many of the industry-based or equipment-based C standards also pertain to manufacturing issues such as equipment dimensions or stability tests.

Technical committee activity can be found through accessing the ISO Web page (www.iso.org). Standards can be easily found by searching under "ergonomics" or else by the Technical Group Ergonomics (TC 159). Under the TC there are four subcommittees that are responsible for different standard developments: general ergonomic principles, anthropometry and biomechanics, ergonomics of human-system interaction, and ergonomics of the physical environment. The standards that have been developed or are under development are accessible through each of the subcommittees. To provide an idea of the types of ISO ergonomics standards available in this section, several of them are grouped by topic in Table 1.

British Standards Institute

The British Standards Institute (BSI) is the primary standards-setting body in the United Kingdom. A number of notable publications of BSI have begun to affect business in the United States, namely the documents that are in the Occupational Health and Safety 18000 series (BSI, 2007, 2008a, 2008b). BSOHSAS 18001:2007 (*Occupational Health and Safety Management Systems. Requirements.*) provides basic requirements (BSI 2007). BS OHSAS 18002:2008, *Occupational Health and Safety Management Systems—Guidelines for the Implementation of OHSAS 18001:2007*, fleshes out the requirements, as the document title suggests. There are examples on how to implement OHSAS 18001, which is designed to allow for integrating into all types and sizes of organizations (BSI 2008a). The third part of the 18000 series, BS 18004:2008, *Guide to Occupational Health and Safety Management*, does more than provide guidance on the requirements of the standard; it also addresses management itself so that a company might achieve success at implementing a safety and health management system (BSI 2008b). The OHSAS 18000 series is compatible with the well-established ISO 9001 (quality management standard) and ISO 14001 (environmental management standard), assuring the integration of a management system. The common thread of the management standards is the use of the continuous improvement cycle PDCA (Plan-Do-Check-Act) that is commonly referred to as Deming's cycle (Podgórski 2006). These management-process standards are relevant to *ANSI/AIHA Z10, Occupational Health and Safety Management Systems*, discussed earlier in the chapter.

International Labour Organization Guidelines

Since 1996, there have been several unsuccessful attempts to initiate an ISO standard on occupational safety and health management systems (Markey 1996, Podgorski 2006, ILO 2008). This led to BSI developing OHSAS 18001, originally in 1999; yet this standard is not an international standard. Since there was no international standard on management systems, the ILO developed guidelines on occupational safety and health management systems (ILO-OSH 2001) in cooperation with the International Occupational Hygiene Association (ILO 2001). These guidelines are designed for both governments and organizations. An ILO update in 2002 indicated the guidelines were being considered for adoption by 18 countries and the ILO is now trying to market them aggressively and in competition with BSI (Johnson 2002, ILO 2008).

The OHSAS 18001 was issued before ILO-OSH 2001, but the two standards are now in direct competition to win over global corporations and be adopted in practice as an international standard. Smaller com-

TABLE 1

Tabulation of Some Existing ISO Ergonomics Standards and Others in Development under the Ergonomics Technical Committee (TC 159)

Subcommittee	Standard	ISO Number
TC 159/SC 1–General ergonomics principles	Ergonomic principles in the design of work systems	ISO 6385:2004
	Ergonomic principles related to mental work-load–General terms and definitions	ISO 10075:1991
	Ergonomic principles related to mental workload–Part 2: Design principles	ISO 10075-2:1996
	Ergonomic principles related to mental workload–Part 3: Principles and requirements concerning methods for measuring and assessing mental workload	ISO 10075-3:2004
Under development:	Ergonomics–General approach, principles and concepts	ISO/DIS 26800
TC 159/SC 3–Anthropometry and biomechanics	Ergonomics–Evaluation of static working postures	ISO 11226:2000
	Ergonomics–Manual handling Part 1: Lifting and carrying Part 2: Pushing and pulling Part 3: Handling of low loads at high frequency	ISO 11228-1:2003 ISO 11228-2:2007 ISO 11228-3:2007
	Ergonomic procedures for the improvement of local muscular workloads Part 1: Guidelines for reducing local muscular workloads	SO/TS 20646-1:2004
	Ergonomic design for the safety of machinery (3 Parts).	ISO 15534-Parts 1–3
TC 159/SC 4–Ergonomics of human-system interaction	Ergonomic requirements for office work with visual display terminals (VDTs) (17 Parts: 8 software, 9 hardware)	ISO.9241, Parts 1–9: General overview requirements of task, posture and layout. Hardware requirements of visual display, colors, keyboard and input devices. Parts 10–17: Software requirements, usability, dialog principles, dialogs of menu, command, form filling, and direct manipulation.
	Ergonomics of human-system interaction. (Multiple parts addressing different areas: 100 series–software; 300 series–electronic visual displays; 400 series–physical input devices and 900 series–tactile and haptic interactions.	ISO.9241 Parts 1xx–4xx, 9xx
	Ergonomic requirements for the design of displays and control actuators (3 parts).	ISO 9355 Parts 1-3
	Ergonomic design of control centres (7 parts).	ISO 11064 Parts 1-7
	Ease of operation of everyday products (4 parts).	ISO 20282 Parts 1-4
TC 159/SC 5–Ergonomics of the physical environment	21 standards pertaining to heat stress and thermal environments. A few are highlighted below:	
	Ergonomics of the thermal environment Risk assessment strategy for the prevention of stress or discomfort in thermal working conditions.	ISO 15265:2004
	Cold workplaces Risk assessment and management	ISO 15743:2008
	Evaluation of thermal environments in vehicles (2 parts).	ISO/TS 14505 Parts 1–2
	Methods for the assessment of human responses to contact with surfaces (3 parts)	ISO 13732- Parts 1-3
	Ergonomics–Assessment of speech communication	ISO 9921:2003
	Ergonomics–System of auditory and visual danger and information signals	ISO 11429:1996
Under development	Ergonomics–Accessible design Auditory signals for consumer products	ISO/FDIS 24500
Under development	Ergonomics–Accessible design Sound pressure levels of auditory signals for consumer products	ISO/DIS 24501
Under development	Ergonomics–Accessible design Specification of age-related luminance contrast in visual signs and displays	ISO/DIS 24502

(*Source*: ISO 2010)

panies in the United States and elsewhere will shortly follow any global lead. The underlying health and safety management system of a company is relevant to ergonomics, and the process of managing ergonomic-related problems is itself an essential component of effective control of such problems (NRC 2001).

CANADIAN ERGONOMICS STANDARDS

British Columbia

In the fall of 1994, the Workers' Compensation Board of British Columbia (BC) issued a draft ergonomics regulation. The regulation failed to be adopted by the BC Legislature in 1995. However, since 1994, there has been a two-page section on ergonomics in "Part 4, General Conditions" of the Occupational Health and Safety Regulations of the Workers' Compensation Board of BC (British Columbia 2006). "Sections 4.46–4.53, Ergonomics (MSI) Requirements," charge employers with identifying factors that might expose workers to the risk of a musculoskeletal injury (MSI), to assess the identified risks, and to eliminate or minimize the risks. Employees are to receive education and training and be consulted by the employers. Evaluation of effectiveness is required. Since this is written into the regulations BC has useful materials on its Web site to support companies in addressing musculoskeletal disorders. Other provinces, such as Manitoba, have adapted these materials in their ergonomics guidelines to improve safety and health (Manitoba Labor and Immigration 1999).

Saskatchewan

The government of Saskatchewan has elements that address ergonomics in sections of The Occupational Health and Safety (OH&S) Regulations of 1996, although the term ergonomics is not used in the regulations. Sections 78–83 cover lifting and handling loads, standing for long periods of time, appropriate jobs for sitting and seating requirements, identifying and controlling musculoskeletal injuries, constant effort and exertion, and visually demanding tasks (Saskatchewan Labor 1996). A Code of Practice for Visual Display Terminals was issued in 2000 that provides guidance on complying with the OH&S regulations at computer workstations.

Ontario

The Ministry of Labour, Government of Ontario (ON), drafted legislation titled Physical Ergonomics Allowable Limits. The report was rescinded and shelved in 1995–96. Ontario's Occupational Health and Safety Act of 1979 was changed in 1990 with some significant additions. All employers have to have a health and safety policy and program and the officers of corporations have direct responsibility. In workplaces of greater than twenty workers there has to be a joint labor and management Health and Safety Committee that is responsible for health and safety in the workplace. The committee is to meet regularly to discuss health and safety concerns, review progress, and make recommendations. Workplaces of less than twenty must have a Health and Safety Representative. By 1995, employers had to certify that the members of their joint health and safety committees were properly trained (BNA 1994).

In February of 2005, the government of Ontario set a goal to reduce workplace injuries by 20 percent by 2008. In order to meet this goal the Ministry of Labour established an ergonomics working panel to make recommendations on how to reduce ergonomic-related injuries, coordinate 14 health and safety associations to develop prevention strategies, provide research monies, allocate funds for bed lifts to reduce back injuries to nurses, and hire more inspectors. The government's approach does not include an ergonomics standard per se (Ontario Ministry of Labor 2005).

Since 2007, the Ontario government has issued a three-part Musculoskeletal Disorders Prevention Series, in partnership with the Occupational Health and Safety Council of Ontario (OHSCO). Part 1, *MSD Prevention Guideline for Ontario*, provides a framework by which to address MSDs. Part 2 is a resource manual for the guideline and Part 3, the *MSD Prevention Toolbox* is divided into Parts A, B, and C (OHSCO 2007, Ontario Ministry of Labor 2010).

Canadian Standards Association

The Canadian Standards Association (CSA) produces voluntary standards pertaining to many areas. A primary standard is CSA Z1000-06, *Occupational Health and Safety Management*, which provides an overall framework for prevention of injuries and illness. It is also compatible with other management systems using the Plan-Do-Check-Act (PDCA) method. The draft standard Z1004, *Workplace Ergonomics*, is under public review. Z1004 is designed to be used with Z1000 but can also be used as an independent document. The draft of *Workplace Ergonomics* delineates the role ergonomics plays in occupational safety and health.

Another guideline that is widely used is CSA-Z412, *Guideline on Office Ergonomics*. The latest version (2000) is produced as an interactive CDROM, Adobe PDF file, or as hardcopy. The guideline is comprehensive and practical, covering office systems, furniture and equipment, the environment, and how to conduct an ergonomic analysis (CSA 2000).

Other International Standards

The main ergonomic-related standards of the United Kingdom, Australia, and Japan are summarized in a chapter in the *Handbook of Standards and Guidelines in Ergonomics and Human Factors* (Stuart-Buttle 2006c). The handbook also has chapters providing detail on other international standards (Karwowski 2006).

Conclusion

The regulatory focus on ergonomics in the United States fluctuates, but overall, since the late 1980s, there has been a considerable volume of activity surrounding regulation. The debate in the nation continues on how best to address the large occupational safety and health problem of musculoskeletal disorders. In the meantime, international standards continue to be developed, and many companies turn to guidelines used in other countries, such as Canada. Useful resources can be found in the Appendix at the end of the chapter.

References

Abrams, A. L. 2002. "OSHA & Ergonomics; Enforcement Under the General Duty Clause." *Professional Safety*, September 2002, pp. 50–52.

American Conference of Governmental Industrial Hygienists (ACGIH). 2001. *2001 TLVs and BEIs: Threshold Limit Values for Chemical Substances and Physical Agents and Biological Exposure Indices*. Cincinnati, Ohio: ACGIH.

American Industrial Hygiene Association (AIHA). 2005. *ANSI/AIHA Z10 Occupational Health and Safety Management Systems*. Fairfax, VA: AIHA.

_____. 2008. *Ergonomics Program Guidance Document*. Rostykus, W., ed. Fairfax VA: AIHA.

American Nurses Association (ANA). 2010. "Safe Patient Handling and Movement (SPHM): Background." State Legislative Agenda, SPHM, www.nursingworld.org.

American Society of Safety Engineers (ASSE). 2009. "Prevention Through Design (PTD): Guidelines of Addressing Occupational Risks in Design and Redesign Processes." Technical Report TR-Z790.001.2009. Des Plaines, IL: ASSE.

_____. 2010. "ASSE Launches Prevention Through Design Initiative, Holds Feb Webinar." *ASSE News*. Jan 11, 2010.

Anderson, J. 2006. "Safe Patient Lifting Legislation Makes Progress." October 9, *ErgoWeb*. www.ergoweb.com

Bureau of National Affairs (BNA). 1991. "Lantos Blasts OSHA for Standards Delay While Workers Suffer from 'National Epidemic'." *Occupational Safety and Health Reporter* (March 27, 1991), p. 1522.

_____. 1995. "Ontario Ministry Moves Up Deadlines for Meeting Certification Requirements." *Occupational Safety and Health Reporter* (March 30, 1995), p. 1430.

_____. 1997. "Ergonomic Hazards Properly Cited Under General Duty Clause, Commission Says." *Occupational Safety and Health Reporter*, April 30, 1997, pp. 1491–1492.

_____. 2001. "State Businesses, National Associations Join Lawsuit on Washington State Rule." *Occupational Safety and Health Reporter* (October 25, 2001), pp. 965–966.

_____. 2002. "Beverly Settlement Raises Complex Issue of Using Law, Not Rule, to Protect Workers." *Occupational Safety and Health Reporter* (January 24, 2002) 32(4):73–75.

_____. 2003. "Safety Council Must Be 'Willing and Able' to Continue as Secretariat, ANSI Panel Says." *Occupational Safety and Health Reporter* (October 16, 2003) 33(41):1000–1001.

_____. 2004a. "Ergonomics Process Falters in States; Defeat, Delay Plague 2003 Standards Efforts." *Occupational Safety and Health Reporter* (January 22, 2004) 34(1):78–79.

_____. 2004b. "ABA Members Concerned Over Ergonomics, Repeat OSHA Violations, Targeted Inspections." *Occupational Safety and Health Reporter* (March 11, 2004) 34(11):267–268.

_____. 2004c. "VPP Challenge, Corporate Pilots Started; OSHA Moving Forward with VPP Growth." *Occupational Safety and Health Reporter* (June 3, 2004) 34(23):567–568.

_____. 2004d. "Construction Voluntary Protection Program Details Published; OSHA Requests Comments." *Occupational Safety and Health Reporter* (September 9, 2004) 34(36):901–902.

_____. 2005. "Michigan Begins Drafting Possible Standard; State Bill Planned to Stop Ergonomic Rules." *Occupational Safety and Health Reporter* (March 3, 2005) 35(9):177.

_____. 2009. "OSHA Proposes Injury, Illness Form Changes, Denies Plans for Broader Ergonomics Standard." *Occupational Safety and Health Reporter* (February 4, 2010) 40(5).

British Columbia (BC) Workers' Compensation Board. 2006. *Occupational Health and Safety Regulations of the Workers' Compensation Board of BC*. British Columbia: WC Board.

British Standards Institute (BSI). 2007. BS OHSAS 18001. *Occupational Health and Safety Management Systems. Requirements*. www.bsiamerica.com/Standards-and-Publications

_____. 2008a. BS OHSAS 18002:2008. *Occupational Health and Safety Management Systems—Guidelines for the Implementation of OHSAS 18001:2007*. www.bsiamerica.com/Standards-and-Publications

_____. 2008b. BS 18004:2008. *Guide to Occupational Health and Safety Management*. www.bsiamerica.com/Standards-and-Publications

California Occupational Safety and Health Administration (CAL/OSHA). 1997. S5110. *Repetitive Motion Injuries*. California Code of Regulations, Title 8, Section 5110. www.dir.ca.gov/scripts/samples/search/query.id

Chapman, C. D. 2006. "Using Kaizen to Improve Safety and Ergonomics." *Occupational Hazards* (February 2006), pp. 27–29.

Canadian Standards Association (CSA). 2000. CSA Z412. *Guideline on Office Ergonomics*. www.csa.ca

_____. 2006. CSA Z1000-06. *Occupational Health and Safety Management*. www.csa.ca

_____. 2010. CSA Z1004, Draft. *Workplace Ergonomics*. www.csa.ca

Dolan, S. 1998. "Health and Safety Initiatives Help Enhance Compaq's Profitability." *Safe Workplace*, Spring, pp. 23–25.

Dwyer, W., and C. Lotz. 2003. "An Ergonomic 'Win-Win' for Manual Material Handling." *Occupational Hazards* (September 2003), pp. 108–112.

Equal Employment Opportunity Commission (EEOC). 1990. *Americans with Disabilities Act of 1990*. Public Law 101-336, 101st Congress (July 26, 1990). Washington, D.C.: U.S. Government Printing Office.

Federal Register. 1989. *Safety and Health Program Management Guidelines*. Federal Register 54 (16) (January 26, 1989). Washington, D.C.: U.S. Government Printing Office.

_____. 2000a. *Occupational Safety and Health Administration: Ergonomics Program; Final Rule*. Federal Register 65 (220) (November 14, 2000). Washington, D.C.: U.S. Government Printing Office.

_____. 2000b. *Occupational Safety and Health Administration: Revisions to the Voluntary Protection Programs to Provide Safe and Healthful Working Conditions; Notice*. Federal Register. Notices. 65 (142) (July 24, 2000). Washington, D.C.: U.S. Government Printing Office.

Fulwiler, R. D. 1998. "People, Public Trust and Profit; Modeling Safety for Lasting Success." *Occupational Hazards* (April 1998), pp. 36–39.

"Gearing Up for OSHA's Ergonomics Guidelines." 1990. *Occupational Hazards* (April 1990), pp. 47–48.

Geddes Lipold, A. 2003. "The Soaring Cost of Workers' Comp." *Workforce Management*. www.workforce.com

General Accounting Office (GAO). 1997. *Worker Protection: Private Sector Ergonomics Programs Yield Positive Results*. Washington, D.C.: U.S. Government Printing Office.

Government Accountability Office (GAO). 2009. *OSHA's Voluntary Protection Programs: Improved Oversight and Controls Would Better Ensure Program Quality*" GAO-09-395 (May 20, 2009). www.gao.gov

Greenhouse, S. 2001. "House Joins Senate in Repealing Rules Issued by Clinton on Work Injuries." *The New York Times*, March 8, 2001.

Hendrick, H. W. 1996. "Good Ergonomics Is Good Economics." Presidential address. Proceedings of the Human Factors and Ergonomics Society 40th Annual Meeting, Santa Monica, CA, pp. 1–15.

Hudson, A. 2007. *Safe Patient Handling, Legislative Update May 2007* (May 13, 2007). Work Injured Nurses' Group (WING). www.wingusa.org

_____. 2010. *WING USA Safe Patient Handling, Update February 28, 2010: "CHAPS" Return to Capitol Hill*. www.wingusa.org

Human Factors and Ergonomics Society (HFES). 2007. ANSI/HFES 100, *Human Factors Engineering of Computer Workstations*. Santa Monica, Calif.: HFES.

_____. 2008. ANSI/HFES 200, *Human Factors Engineering of Software User Interfaces*. Santa Monica, Calif.: HFES.

Hogrebe, M. C. 2004. "Video Microscopes for our Video Generation." *ErgoSolutions Magazine* (February 29, 2004), pp. 22–24.

Industrial Safety and Health News (ISHN). 2009. "VPP's Star Dims In Obama Administration." Editorial June 26. *Industrial Safety and Health News*. www.ishn.com

International Labor Organization (ILO). 2001. *Guidelines on Occupational Health and Safety Management Systems*. ILO/OSH 2001. Geneva, Switzerland: ILO. www.ilo.org

_____. 2008. *Report of the Director-General: Seventh Supplementary Report*. GB.301/17/7 301st Session. www.ilo.org

International Standards Organization (ISO). 2010. Technical Committee (TC) 159 Ergonomics. www.iso.org/iso/standards_development/technical_committees/list_of_iso_technical_committees/iso_technical_committee.htm?commid=53348

Johnson, D. 2002. "Are We Getting Closer to a Global Safety and Health Standard? The Race to Fill the Void." *The Synergist* (August 2002), pp. 24–27.

Karr, A. R. 2002. "Ergonomics 101: Business Resists Course." *Wall Street Journal*, January 26, 2002.

Karwowski, W., ed. 2006. *Handbook of Standards and Guidelines in Ergonomics and Human Factors*. New Jersey: Lawrence Erlbaum.

Kirchgaessner, S. 2010. "U.S. Business Opposes Work Safety Proposal." *Financial Times* (February 2). www.FT.com

Laws, J. 2010. "Michigan's Ergonomics War Set to Resume." *Occupational Health and Safety* (February). Online at www.ohsonline.com

Maloney, D. 2000. "The Ergonomic Silver Lining." *Modern Materials Handling* (February 2000), pp. 53–58.

Manitoba Labor and Immigration. 1999. *Manitoba's Ergonomics Guideline*. Workplace Safety and Health Division, Manitoba, Canada. www.gov.mb.ca/labour/safety/ergoguide.html

Manuele, F. 2006. "ANSI/AIHA Z10-2005, The New Benchmark for Safety Management Systems." *Professional Safety* (February 2006), pp. 25–33.

Marras, W. S., and C. Hamrick. 2006. The ACGIH TLV® for Low Back Risk. In W. S. Marras and W. Karwowski, eds. *Occupational Ergonomics Handbook*. 2d ed. "Fundamentals and Assessment Tools for Occupational Ergonomics," Ch 50. Boca Raton, FL: CRC Press.

Maygar, S. V. 1999. "Medical Claim Management; Controlling Workers' Compensation Losses." *Professional Safety* (March 1999), pp. 41–45.

McDermott, H., K. Lopez, and B. Weiss. 2004. "Computer Ergonomics Programs." *Professional Safety* (June 2004), pp. 34–39.

Minter, S. G. 2001. "Attack on TLV®: The American Conference of Governmental Industrial Hygienists Enters a High-Stakes Legal Battle Over Its Most Famous Product—TLVs. *Occupational Hazards* (May 2001), p. 6.

Mugno, S. A. 2002. "Voluntary Efforts Derail Ergonomic Injuries." *Safety + Health* (February 2002), pp. 25, 27.

National Institute of Standards and Technology (NIST). 2010. *Eighty-three Organizations Seek the 2010 Baldrige Award* (June 1, 2010). www.nist.gov

National Research Council (NRC) and the Institute of Medicine. 2001. *Musculoskeletal Disorders and the Workplace: Low Back and Upper Extremities*. Panel on Musculoskeletal Disorders and the Workplace. Commission on Behavioral and Social Sciences and Education. Washington, D.C.: National Academy Press.

Occupational Health and Safety Council of Ontario (OHSCO). 2007. Musculoskeletal Disorders Prevention Series: Part 1, *MSD Prevention Guideline for Ontario*. Toronto, ON, Canada: OHSCO. www.esao.on.ca/downloads/MSD.aspx

_____. 2007. Musculoskeletal Disorders Prevention Series: Part 2, *Resource Manual for the MSD Prevention Guideline for Ontario*. Toronto, ON, Canada: OHSCO. www.esao.on.ca/downloads/MSD.aspx

_____. 2008. Musculoskeletal Disorders Prevention Series: Part 3, *MSD Prevention Toolbox* (Parts A, B, and C). Toronto, ON, Canada: OHSCO. www.esao.on.ca/downloads/MSD.aspx

Occupational Safety and Health Administration (OSHA). 1970. *Occupational Safety and Health Act of 1970*. Public Law 91-596, 91st Congress, S. 2193, December 29, 1970. As amended by Public Law 101-552, 3101, November 5, 1990. Washington, D.C.: U.S. Government Printing Office.

_____. 1990. OSHA 3123. *Ergonomics Program Management Guidelines for Meatpacking Plants*. Washington, D.C.: OSHA.

_____. 2002a. "OSHA announces Comprehensive Plan to Reduce Ergonomic Injuries." OSHA National News Release. USDL 02-201, April 5, 2002. www.osha.gov/media/oshnews/apr02/national-20020405.html

_____. 2002b. *A Four-Pronged, Comprehensive Approach*. April 8, 2002. www.osha.gov/ergonomics/ergofact02.html

_____. 2005a. *Partnership: An OSHA Cooperative Program*. April 30, 2005. www.osha.gov/dcsp/partnerships/index.html

_____. 2005b. *OSHA's Safety and Health Achievement Recognition Program (SHARP)*. April 30, 2005.

_____. 2009a. *Site-Specific Targeting 2009*. SST-09. Directive number: 09-05 (CPL 02). July 20, 2009. www.osha.gov

_____. 2009b. "U.S. Department of Labor's OSHA Revises its Voluntary Protection Programs." Trade news release (January 9, 2009).

_____. 2010. *OSHA Statistics and Data. Inspection Data*. www.osha.gov/oshstats/www.osha.gov/dcsp/small-business/sharp.html

Ontario Ministry of Labour. 2005. *Repetitive Strain Injuries and Ergonomics*. Backgrounder Document d'information, February 28, 2005. www.gov.on.ca

"OSHA Seeks VPP Growth for Worker Protection." 2000. *Safety + Health* (November 2000), p. 18.

Oxenburgh, M., and P. Marlow. 2006. "Cost Justification for Implementing Ergonomics Intervention." In Marras, W. S., and W. Karwowski, eds. *Occupational Ergonomics Handbook*. 2d ed. "Interventions, Controls, and Applications in Occupational Ergonomics," Ch 4. Boca Raton, FL: CRC Press.

Podgórski, D. 2006. "ILO Guidelines on Occupational Safety and Health Management Systems." In W. Karwowski, ed. *Handbook of Standards and Guidelines in Ergonomics and Human Factors*. Ch 27, pp.493–505. Mahwah, NJ: Lawrence Erlbaum Assoc.

Public Law. 2006. *National Technology Transfer and Advancement Act of 1995*. Public Law 104-113, 104th Congress. 110 Stat.775. Washington, D.C.: U.S. Government Printing Office.

Risk and Insurance Online. 2009. *Michigan: Senator Introduces Bill to Block State-Mandated Ergonomics Standard* (March 23, 2009). www.RiskandInsurance.com

Rodrigues, C. 2001. "Ergonomics to the Rescue." *Professional Safety* (April 2001) pp. 32–34.

Safetyworks. 1989. *Title 26: Labor and Industry*. Chapter 5, "Health and Safety Regulations." Subchapter 2-A: Video Display Terminal Operators [Heading: PL 1989, c. 512 (new)] §251 and 252. Maine Department of Labor. janus.state.me.us/legis/statutes/26/title26 sec251.html

Saskatchewan Labour. 1996. *The Occupational Health and Safety Regulations, 1996*. Saskatchewan, Canada. www.labour.gov.sk.ca/safety/fast/ergonomics

Sherehiy, B., D. Rodrick, and W. Karwowski. 2006. Chapter 1, "An Overview of International Standardization Efforts in Human Factors and Ergonomics." Karwowski, W., ed. *Handbook of Standards and Guidelines in Ergonomics and Human Factors*. New Jersey: Lawrence Erlbaum.

Smitha, M. T., K. R. Oestenstad, and K. C. Brown. 2001. "State Workers' Compensation, Reform and Workplace Safety Regulations." *Professional Safety* (December 2001), pp. 45–50.

Stuart-Buttle, C. 2006a. Injury Surveillance Database Systems. In Marras, W. S., and W. Karwowski, eds. *Occupational Ergonomics Handbook*. 2d ed. "Interventions, Controls, and Applications in Occupational Ergonomics," Ch 6. Boca Raton, FL: CRC Press.

_____. 2006b. Ergonomics Process in Small Industry. In Marras, W. S., and W. Karwowski, eds. *Occupational Ergonomics Handbook*. 2d ed. "Interventions, Controls, and Applications in Occupational Ergonomics," Ch 8. Boca Raton, FL: CRC Press.

_____. 2006c. Chapter 5, "Overview of National and International Standards and Guidelines." Karwowski, W., ed. *Handbook of Standards and Guidelines in Ergonomics and Human Factors*. New Jersey: Lawrence Erlbaum.

Treaster, J. B. 2003. "Cost of Insurance for Work Injuries Soars Across U.S." *The New York Times*, June 23, 2003.

Walter, L. 2010. "OSHA Introduces New Strategic Plan." *EHS Today* (August 8, 2010).

Washington Department of Labor and Industries. 2006a. *Hazard Prevention; Ergonomics; Rule History*. www.lni.wa.gov/Safety/Topics/Ergonomics/History/default.asp

_____. 2006b. *Hazard Prevention; Ergonomics*. www.lni.wa.gov/Safety/ Topics/Ergonomics/default.asp

APPENDIX: INTERNET RESOURCES FOR REGULATIONS

Federal and National Resources

www.acgih.org   American Conference of Governmental Industrial Hygienists

www.aiha.org   American Industrial Hygiene Association

www.ansi.org   American National Standards Institute

www.asme.org   American Society of Mechanical Engineers

www.asse.org   American Society of Safety Engineers

www.astm.org   American Society for Testing and Materials

www.gpo.gov   General Printing Office

www.hfes.org   Human Factors and Ergonomics Society

www.ieee.org   Institute of Electrical and Electronics Engineers

http://standards.nasa.gov   NASA Technical Standards Program

www.nist.gov   National Institute of Standards and Technology

www.nsc.org   National Safety Council

www.osha.gov   Occupational Safety and Health Administration

www.sae.org   Society of Automotive Engineers

State Resources

www.dir.ca.gov   California Ergonomics Standard

www.safetyworksmaine.com   Maine Department of Labor Outreach Program

www.lni.wa.gov/wisha   Washington State, Department of Labor and Industries

International Resources

www.worksafebc.com   British Columbia (BC) Government, Canada

www.bsigroup.com   British Standards Institute, UK

www.csa.ca   Canadian Standards Association, Canada

www.cen.eu   European Committee for Standardization (CEN)

www.europa.eu   European Union (EU)

www.iea.cc   International Ergonomics Association (IEA)

www.ilo.org   International Labour Organization (ILO)

www.iso.org   International Standard Organization (ISO)

www.useuosh.org   Occupational Safety and Health Administration (USA) and European Union (EU)

www.perinorm.com   Perinorm, private database of international, European and national standards

www.worksafebc.com   Workers' Compensation Board of British Columbia, Canada

www.gksoft.com/govt/en/world.html   Worldwide Governments on the WWW

Principles of Ergonomics

Magdy Akladios

2

LEARNING OBJECTIVES

- Gain an overview of the history of ergonomics.
- Understand the basics of ergonomics.
- Recognize the different risk factors that contribute to an ergonomic injury.
- Distinguish between cumulative trauma disorders (CTDs) and musculoskeletal disorders (MSDs).
- Be able to recognize the various types of ergonomic injuries.
- Learn about biomechanics.
- Grasp the basics of how to analyze and measure potential risk factors.
- Understand how the lifting formula (single task) designated by the National Institute for Occupational Safety & Health (NIOSH) is used.

THE TERM ERGONOMICS stems from the Greek words *ergos*, meaning *work*, and *nomos*, meaning *laws*. Put together, it translates to the study of the *laws of work*. It is defined as the science that tries to adapt a task to conform to the capability of the human performing that task. Briefly put, ergonomics aims to fit the task to the human, rather than vice versa. Historically, ergonomics has been used synonymously with *human factors*. However, more and more they are becoming two distinct and unique sciences. While ergonomics deals with studying the physiological effects of work activities on people, human factors deals with the interaction between human beings and their work environment, including machines and other systems. Ergonomics aims at designing machines to reduce injuries to the humans who operate them, while human factors aims at reducing human error through the understanding of psychology, sensory input, error, and noise—in terms of outside interferences and factors (Eastman Kodak Company 1983).

The International Ergonomics Association (IEA 2000) adopted an official definition of ergonomics as follows:

> Ergonomics (or human factors) is the scientific discipline concerned with the understanding of interactions among humans and other elements of a system, and the profession that applies theory, principles, data and methods to design in order to optimize human well-being and overall system performance.

HISTORY OF ERGONOMICS

Ergonomics is said to have started in ancient history. Imhotep (2667–2648 B.C.), the Egyptian chief architect, physician, and high priest of Heliopolis, was credited with the first report of documenting the treatment of back pain resulting from the Egyptians' work habits during the building of King Djoser's (2687–2668 B.C.) step pyramid at Saqqara.

The next known report came thousands of years later, when the physician Bernardino Ramazzini (1633–1714) wrote about the work-related complaints he saw during his medical practice in 1713. His publication was called *De Morbis Artificum*, or *The Diseases of Workers*. However, up to that point, this science had no known or recognizable label.

It was in 1857 that Wojciech Jastrzebowski, a Polish scholar and professor of natural sciences at the Agronomical Institute in Warsaw-Marymount, coined the term ergonomics.

While the industrial revolution was taking hold in all manufacturing fields, many tasks were still performed by hand. In the early 1900s, a number of accommodations were developed to improve worker productivity and efficiency. At the time, ergonomics took a turn toward management, and was dubbed *Scientific Management*. In 1911, Frederick Winslow Taylor, a management consultant, developed the "One Best Way" method to enhance worker production. Taylor was ranked at the top, along with Darwin and Freud, as one of the daring and forward thinkers of modern times. Frank and Lillian Moller Gilbreth pioneered the science of time-motion studies and were able, through job evaluations and analysis, to reduce the time and effort by which a task could be conducted. After her husband's death, Lillian continued her work in the science of Scientific Management and was dubbed "The Mother of Management." The Gilbreths' efforts reduced the number of motions in bricklaying from 18 down to 4.5, hence enhancing productivity from 120 to 350 bricks per hour.

World War II (1939–1945) marked greater sophistication in military equipment and, subsequently, in nonmilitary applications. As a result, human-machine interaction captured greater interest, especially in airplane and cockpit design. This paved the way for such human factors designs to occupy center stage during the race for outer space supremacy during the late 1960s and early 1970s.

As the interest in ergonomics and human factors engineering grew, supporting societies and organizations started to sprout up around the world. The Ergonomics Research Society was first formed in Great Britain in 1949. This society started the *Ergonomics Journal* in 1957 (The Ergonomics Society 2007). In the United States, the Human Factors Society was founded in 1957, and has now grown to 22 technical groups and numerous local and student chapters. Known today as the Human Factors and Ergonomics Society (www.hfes.org), it publishes the journal *Human Factors*.

In 1961, the first International Ergonomics Association meeting was held in Stockholm, Sweden. This organization has members from the United States, the United Kingdom, Japan, Australia, Scandinavia, and other countries interested in that type of research.

In 1990, the Board of Certified Professional Ergonomists (BCPE) (www.bcpe.org) was established to certify human factors and ergonomics practitioners, researchers, and engineers in the United States.

USES AND BENEFITS OF ERGONOMICS

Ergonomics recognizes that every human-machine system involves the interaction between humans and machines in an environment where information is being transferred back and forth between the two, with the human being in control, and the machine providing adequate feedback to the human. Hence, the science of ergonomics, typically, is a combination of a variety of sciences, including (Wickens, Gordon, and Liu 1998):

1. Physiology, which involves studying the effects of work on heart rate, oxygen consumption, perspiration, calories, extreme temperature stresses, and so on.
2. Anatomy, which involves studying the anatomical structure of the different body parts, their relationship to each other, how the human body works in general, and the effects of work on these body parts.
3. Psychology and human factors engineering, which involves the decision-making process of human beings, especially in critical situations, and recognition of human limitations in terms of visual, auditory, and other sensory capabilities.
4. Anthropometry, which is the study of physical dimensions in people, including the measurement of human body characteristics such as

strength, size, breadth, girth, and distance between anatomical points. Anthropometry also includes segment masses, the centers of gravity of body segments, and the ranges of joint motion, which are used in biomechanical analyses of work postures. These measurements are mostly useful for the design of things that are used by people, such as door heights, force needed to turn door knobs, and so on.

5. Biomechanics, which involves measuring forces, angles, loads, and the study of muscular activity (see a more detailed definition of biomechanics in this chapter).
6. Engineering, which considers the previous five sciences, uses engineering design concepts, and applies scholarly thinking to design better tools, machines, equipment, controls, and control rooms. Ergonomists typically find themselves working with a variety of engineers, including aeronautical, computer, information systems, nuclear, civil, and architecture.

As a result, the ultimate aim of ergonomics is to anticipate (and hence reduce and possibly eliminate) potential injuries, accommodate and enhance human performance, and provide an environment where humans and machines work seamlessly and in harmony with each other. This system ultimately enhances productivity; reduces claim costs and pain and suffering; and provides workers with life-long enjoyment at work—hence self-achievement and worth (Sanders and McCormick 1993).

RECOGNIZING POTENTIAL ERGONOMIC PROBLEMS THROUGH THE ANATOMY AND PHYSIOLOGY OF SPECIFIC DISORDERS

Defining Cumulative Trauma Disorders (CTDs) and Associated Risk Factors

Ergonomic problems arise from minute injuries (microtrauma) to the musculoskeleton due to exposures to certain factors. When these exposures are repeated (repeated trauma) due to work habits and environment, they gradually develop into what is now known as cumulative trauma disorders (CTDs), also known as repetitive-motion injuries (Putz-Anderson 1988).

CTDs were recognized in the early days of ergonomics and over the years by the trade with which they were associated. Names like bricklayer's shoulder, carpenter's elbow, stitcher's wrist, gamekeeper's thumb, housemaid's knee, telegraphers' cramp, and cotton twister's hand were recognized in the medical field as disorders that were caused by the type of work with which patients were involved. It was later that the factors that contribute to these disorders were recognized as what is known today as risk factors. While these factors may not be the sole cause of a CTD, the existence of one or more of them will enhance the probability that a CTD gradually develops.

These risk factors include:

- forceful muscle exertion
- repetitive motion
- awkward postures
- vibration
- contact stresses

Risk factors may also include exposure to extreme temperatures, and/or lack of rest to allow the body enough recovery time.

Types of Ergonomic-Related Injuries

Ergonomic-related injuries are either chronic (hence called cumulative trauma disorders), or acute, as a result of a blow, cut, or fall. These result in strains (injuring muscles and tendons) and/or sprains (injuring ligaments) (Putz-Anderson 1988).

The following paragraphs describe the different types of injuries and the different body structures that are affected by them.

Muscle Injuries: Muscles are made up of tiny fibers laid out in the same direction, and are filled with blood vessels to provide oxygen to the muscles and carry carbon dioxide and other waste products away from the muscles. Muscles can be injured in one of three ways: (1) muscle fiber can be strained or irritated; (2) tiny muscle fiber can be torn due to excessive use; and (3) muscle fiber can be crushed due to a severe blow (Putz-Anderson 1988).

Tendon Injuries: Tendons are ropelike structures that connect muscles to bones. Their job is to transfer the motion from the muscle to the bone. Like muscles, tendons can be torn like frayed rope or, because of use, can be inflamed, causing pain, such as tendonitis or inflammation of the tendons (*itis* is a Greek word meaning *inflammation of*). Tendon disorders of the upper limb can include tendonitis, tenosynovitis, stenosis tenosynovitis, trigger finger, De Quervain's disease, ganglionic cyst, unsheathed tendons, golfer's elbow, tennis elbow, and rotator cuff tendonitis (Putz-Anderson 1988).

Ligament Injuries: Ligaments are strong, ropelike fibers that connect bones to each other. Where there are joints, there are ligaments to bind bones together and to limit the range of motion. Injuries to ligaments can include torn fibers, or fibers that tear loose from the bones to which they are attached. Sometimes they can even be torn completely from the bone. This typically occurs due to a sudden strong blow to that area. Because ligaments have poor blood supply, it might take months for torn ligaments to heal (Putz-Anderson 1988).

Bursa Injuries: There are 160 bursae in the body. These are fluid-filled sacs that are located prevalently where there is potential for rubbing pressure due to motion of a tendon or a ligament on top of a bone. Their function is to act as a lubricant to facilitate and ease the motion of these ligaments, tendons, and bones against each other. If bursae did not exist where they are now, continuous rubbing action of tendons on bones will cause inflammation to the tendons, and will soon cause tears in these tendons/muscles/ligaments. Injuries to bursae can include torn sacs, leaky fluids, inflamed bursae, or hardened sacs. Injuries can also include sudden strong blows to the bursae, which can cause traumatic rupture of the sac (Putz-Anderson 1988).

Nerve disorders: As the name implies, nerve disorders are those that involve pressure on a nerve. Pressure can be applied by poorly designed tool handles, equipment, hard work surfaces, or nearby body parts that have been overworked or swollen, such as inflamed tendons, ligaments, or bones. Examples can include carpal tunnel syndrome and trigger finger. Pain and tingling sensations are similar to hitting the "funny bone" of the elbow (Putz-Anderson 1988).

Neurovascular disorders: These are disorders that involve some combination of nerves (*neuro*), and blood vessels (*vascular*). A prime example is when blood vessels between the neck and the shoulder are compressed, causing a limitation to the amount of blood supply to that area of the body. This compression can be caused by applying pressure directly to that area by a tool or equipment, or by continuously abducting and turning the shoulders, causing muscles of that area to apply the pressure. When blood supply (and the oxygen, nutrients, and other materials the blood carries) is limited, the recovery of damaged muscles and other parts is impeded or, at best, slowed down considerably. Examples can include thoracic outlet syndrome and a variety of vibration-related syndromes (Putz-Anderson 1988).

SOME COMMON DISORDERS

As mentioned earlier, there are a large number of injuries that can occur as a result of exposure to risk factors. However, to limit the scope, only some of the most prevalent injuries will be discussed. These include:

- carpal tunnel syndrome
- trigger finger
- low-back pain
- thoracic outlet syndrome
- cubital outlet syndrome

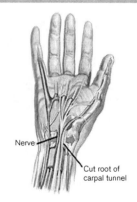

FIGURE 1. Carpal tunnel anatomy and surgery (*Source:* American Academy of Orthopedic Surgeons 2005)

Carpal Tunnel Syndrome

Carpal tunnel syndrome (CTS) is responsible for an average of 25 lost work days per case, resulting in an enormous amount of losses due to workers' compensation claims. Due to their smaller structure and hormones, females are at a higher risk of getting CTS than are males. Also, pregnancy causes a shift in female hormonal balance leading to inflammation as well.

Anatomy of CTS

CTS occurs when the median nerve, which passes through a tunnel that is located in the wrist area of the hand, is under pressure. The bottom of that tunnel is made out of tendons that are encased in tendon sheaths and the transverse carpal ligament from the top. That ligament is as hard as bone and has no elasticity.

Causes of CTS

When an individual is exposed to awkward postures, excessive muscular motion of the hands and fingers, and other factors, the tendon sheaths on the bottom of the tunnel will get inflamed. This inflammation puts pressure on the median nerve which, in turn, is translated into symptoms of the CTS.

Symptoms of CTS

Symptoms can include numbness, tingling, atrophy, and general weakness in the opposition motion. These symptoms occur at the wrist area where the median nerve is located under the transverse carpal ligament, and radiates to where that nerve flows to from this point. This includes the thumb, the pointer finger, the middle finger, and the inside half of the ring finger (see Figure 1). Pain is mostly excruciating at night, as well as during exposure to the cause of the injury in the first place. Diagnosing CTS includes a medical examination, and in some cases, nerve conduction velocity testing, which rules out subjective diagnosis.

Treatment of CTS

Treatment of CTS includes rest and the wearing of braces to maintain the wrist in a neutral position, high doses of vitamin B6, aspirin or ibuprofen, which tend to reduce the inflammation of the tendon sheaths. Noninflammatory medicines are then administered. A cortisone injection can be administered to aid in the diagnosis process and to provide temporary relief of the symptoms. As a last resort, carpal tunnel release surgery or endoscopic carpal tunnel release surgery is conducted to cut open the transverse carpal ligament to release the pressure off of the median nerve. The body will then close that cut ligament by building scar tissue (Sechrest 1996a).

Trigger Finger

The tendons that control the motion of the fingers are held in place parallel to the finger via ligaments (pulleys). These pulleys are attached to the bones of the fingers and encase the tendons (see Figure 2).

Causes of Trigger Finger

As the name implies, trigger finger results from excessive and forceful use of the finger. Jobs such as spray painters and rivet gun operators are prime candidates for it. The excessive use of the finger results in inflammation of the sheath that encases the tendon. Since these ligaments are hard and have no give, the tendon sometimes gets too large to pass underneath the pulley. In worst-case scenarios, the tendon forms a bulge that prevents the tendon from gliding underneath the pulley. Similar to carpal tunnel syndrome, trigger finger is more common in women than men due to the smaller female structure. It occurs most frequently in people who are between the ages of 40 to 60 (NIOSH 1997).

FIGURE 2. Trigger finger/thumb (*Source:* Medical Multimedia Group LLC 2001)

Symptoms of Trigger Finger

The symptoms include pain and numbness in that area, which in turn causes restricted motion of the finger. Forceful motion of the finger will result in excruciating pain. Diagnosis requires a physical exam. An occupational physician may also force the finger open. In doing so, the bulging tendon will glide underneath the pulley with an audible click and a lot of pain. The click is always a true sign of trigger finger.

Treatment of Trigger Finger

Like carpal tunnel syndrome, treatment might include braces, noninflammatory medicines, and in the worst-case scenario, surgery. During surgery, the ligament (pulley) is cut open to provide the inflamed tendon enough room to pass, thus eliminating the pressure and allowing the inflammation to subside. The body will then close that gap by building up scar tissue (Patel 1997c).

Back Problems

There are around 30,000 occupational back injuries per year, and 80 percent of the adult population in the United States complains of low-back pain. Back problems cost the highest as compared to other nonfatal occupational injuries. Back pain can be caused by: (1) irritation of the large nerve roots that go to the legs and arms; (2) irritation of the smaller nerves that innervate the spine; (3) strain of the large paired back muscles (*erector spinae*); (4) injury of the bones, ligaments, or joints themselves; or (5) the disk space itself can be a source of pain due to an injury such as a herniated disk (Spine-Health.com 2005).

These injuries may be caused by trauma; trips and/or falls; aging-related, degenerative diseases of the bones; herniated disks; and/or lifting (either cumulative or abrupt). Lifting does not necessarily have to involve heavy objects to result in a back problem. In 1981, NIOSH developed a formula to be used as a guideline for jobs that require manual lifting as part of the day-to-day activities (NIOSH 1997). As specified by the NIOSH lifting formula (discussed in more detail in a later section of this chapter), factors that

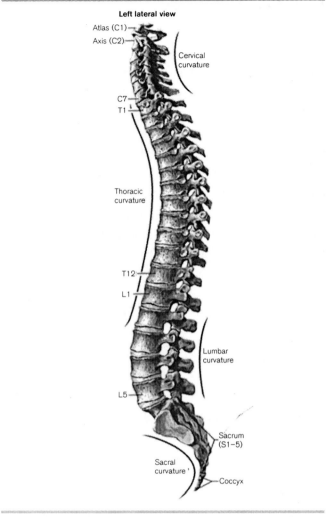

FIGURE 3. Spinal column (*Source:* Netter 1997)

cause back problems due to lifting include the weight of the load, the angle of twist while lifting, the distance away from the body of the lifter, the frequency of lifting, the vertical distance of the lift, and the coupling of the load.

Anatomy of the Back

The back consists of 24 vertebrae stacked on top of each other like a tower, with the thicker vertebrae on the bottom, and the finer, smaller vertebrae toward the top. When standing erect, a healthy spinal column should form a letter "S." The curves of the spine are to facilitate coupled motions of the back, including bending and squatting. These curvatures are called the *lordotic* and *kyphotic* curves (Meyer 2001).

There are four sections to the back: (1) the cervical section, which contains the seven vertebrae that make up the neck or upper back; (2) the thoracic vertebrae, which contain the twelve vertebrae that make up the middle back; (3) the lumbar vertebrae, which contain the five vertebrae that make up the lower back; and (4) the sacrum region. There are intervertebral disks separating the vertebrae (see Figure 3). Their function is to provide cushioning and separation between the vertebrae to prevent excessive rubbing action between vertebrae. Intervertebral disks are made out of an exterior annulus fibrosus, which is a cartridgelike shell, and an interior nucleus pulposus, which is a viscouslike paste.

Causes of Back Injury Due to Lifting

There are many scenarios by which a person can injure his or her back. One of the most common scenarios is as follows: A person who is standing up straight puts equal loads across the surface of the intervertebral disks. However, when this person bends at the back to lift an object, the load is shifted toward the front of the disk, and the rear of the disk is subjected to minimal forces. The resulting compression and sheer force can be enough to squeeze the interior pastelike material (the *nucleus pulposus*) out of the exterior shell (the *annulus fibrosus*) and push against the neighboring nerves. The pressure that these nerves are subjected to translates into pain. This breakage of the shell is commonly known as a *herniated disk*.

Treatment of Back Pain

Most back cases will heal with some physical therapy treatment sessions, if the pain resulted from causes other than a herniated disk. However, herniated disks are sometimes injected with cortisone to help dissolve the nucleus pulposus that is pushing against the nerves. This involves a series of two injections, six months apart. In a worst-case scenario, a procedure is conducted to surgically remove the cause of the pressure, sometimes removing the entire disk, and fusing the upper and lower vertebrae together. The success rate of this surgery is 75–80 percent. The procedure causes some range-of-motion disabilities and creates a fertile ground for further discomfort to the disks above and below the fused vertebrae.

A new treatment, developed and used in Europe and South Africa, calls for replacing the damaged disk(s) with an artificial material. This new procedure is now under extensive research in the United States in a number of large cities, such as New York, Los Angeles, and Chicago (Spinal Motion, Inc. 2005).

Thoracic Outlet Syndrome

The thoracic outlet is a space between the rib cage (*thorax*) and the collar bone (*clavicle*), through which the main blood vessels and nerves pass from the neck and thorax into the arm.

Causes of Thoracic Outlet Syndrome (TOS)

TOS is a neurovascular disorder that is caused by exposing the nerves and blood vessels that pass from the neck and thorax into the arm to excessive chronic pressure. The resulting symptoms are known as thoracic outlet syndrome. In some cases, the cause of compression is a reduction in the space through which these blood vessels and arteries pass (see Figure 4). This reduction in space and compression on the nerves and arteries can be due to repetitive activities that require the arms to be held overhead.

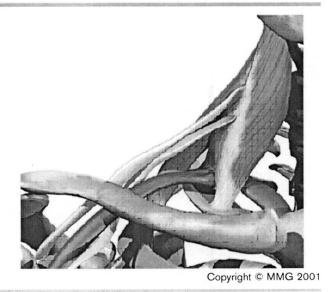

FIGURE 4. Thoracic outlet syndrome
(*Source:* Medical Multimedia Group LLC 2001) 2001

Symptoms of TOS

Symptoms include a combination of pain, numbness, tingling (pressure on sensory nerves), weakness and fatigue (pressure on motor nerves), and/or swelling and coldness in the arm and hand (pressure on blood vessels). The symptoms can mimic many other conditions, such as a herniated disk in the neck, carpal tunnel syndrome, and even bursitis of the shoulder. As a result, TOS can be very difficult to diagnose.

Treatment of TOS

The length of time the arms are used in outstretched or overhead positions should be reduced and spaced out. Taking frequent breaks and changing positions are useful to prevent and eliminate TOS (Patel 1997b). If the symptoms persist, physical therapy and stretching may help alleviate the pressure. When all else fails, surgery for thoracic outlet syndrome is used to expand the tight space that caused the symptoms in the first place. This could be removing an extra rib, or removing the causes of the compression (Sechrest 1996b).

Cubital Tunnel Syndrome

The cubital tunnel is a tunnel that is located behind the funny bone (*medial epicondyle*) on the inside of the elbow. This tunnel is formed by the bone surrounded by muscles and ligaments. The ulnar nerve passes through the cubital tunnel on its way from the arm to the forearm and hand. The ulnar nerve runs to the little finger and the outside half of the ring finger (Eaton 1997).

Causes of Cubital Tunnel Syndrome

In normal subjects, bending the elbow causes the nerve to stretch several millimeters. When this is done repeatedly for activities that require repeated bending and straightening of the elbow in the workplace, the nerve becomes irritated and inflamed. In other patients, the nerve shifts and actually snaps over the prominence of the medial epicondyle. This snapping motion stretches and irritates the nerve. Leaning on the elbow, or resting the elbow on an elbow rest during a long-distance drive or while running machinery can cause repetitive pressure and irritation on the nerve. A direct hit on the tunnel can damage the ulnar nerve. When the ulnar nerve is exposed to pressure or is irritated at the elbow area, numbness and tingling is felt at the elbow area, and runs down the little finger and the ring finger of the hand.

Symptoms of Cubital Tunnel Syndrome

Early signs are numbness on the inside of the hand and in the ring and little fingers. Later there is weakness of the hand. There may be pain at the elbow. Tapping on the nerve as it passes through the cubital tunnel causes tingling or electric shock sensation down to the little finger.

Treatment of Cubital Tunnel Syndrome

The early symptoms of cubital tunnel syndrome usually respond to stopping the activity that is causing the symptoms. This is typically done by reducing the tasks that require repeated bending and straightening of the elbow. Frequent breaks from work for at least five minutes every half hour should be taken. In some patients, the symptoms might be worse at night because they sleep with their elbow bent. If the symptoms fail to respond to activity modifications, surgery might be required to stop progression of damage to the ulnar nerve. The operation moves the ulnar nerve from behind the medial epicondyle to the front of the medial epicondyle. This gives the nerve some slack and removes the stretching of the nerve (Patel 1997a).

OCCUPATIONAL BIOMECHANICS

Biomechanics is a branch of the broader field of ergonomics. While ergonomics attempts to fit the task to the worker, biomechanics, according to Frankel and Nordin, "... uses laws of physics and engineering concepts to describe motion undergone by the various body segments and the forces acting on these body parts during normal daily activities" (Chaffin, Andersson, and Martin 1999).

Based on Chaffin, Andersson, and Martin (1999), occupational biomechanics is defined as "the study of the physical interaction of workers with their tools, machines, and materials so as to enhance worker performance while minimizing the risk of musculoskeletal disorders (MSDs)."

Simply put, biomechanics is the science that studies the effects of internal and external forces on the human body in movement and rest.

Chaffin defines two types of biomechanical injury mechanisms that are common to industry: (1) sudden force as a result of impact trauma, which has sudden outcomes such as amputations, fractures, contusions, and so on, and (2) volitional activity as a result of overexertion and CTD trauma, which has injuries that are cumulative over time, such as tendonitis, nerve entrapments, and other types of MSDs (Chaffin, Andersson, and Martin 1999). Occupational biomechanics is mostly interested in the latter type of activity.

History of Biomechanics

Biomechanics had early beginnings. Many early scientists developed systems to measure and count heart rate and other functions of the body. Leonardo da Vinci developed a great understanding of the functions of muscles and bones, and the musculoskeletal system in general. In the late 1500s, the physicist Galileo Galilei used a system of pendulum oscillations to simulate heart rates. William Harvey, in 1615, demonstrated the importance of capillaries in connecting veins and arteries. In the 1700s, Stephen Hales discovered the elastic functions of the aorta and its importance in producing a smooth flow from the heart to the body.

During the Industrial Revolution in the early 1900s, it was rationalized that the cost of a laborer was cheap, and that it was easier to replace a laborer if injured or killed than it was to fix a workstation. The twentieth century saw great progress in the field of biomechanics in terms of new discoveries, the development of new systems, programs, and standards in developed countries to reduce the pain and suffering experienced by workers. Today, the field of biomechanics uses many other fields of study to minimize the risk of musculoskeletal disorders (Chaffin, Andersson, and Martin 1999).

Methodologies in Biomechanics

As a result of the need for biomechanical analysis and studies, a variety of technologies and sciences started to develop. These include kinesiology, which includes kinematics and kinetics; biomechanical modeling; anthropometry; bioinstrumentation; motion classification and time prediction systems; and mechanical work capacities.

Kinesiology is the scientific study of human movement and the movements of implements or equipment that a person might use in a variety of forms of physical activity. Kinesiology includes kinematics and kinetics. *Kinematics* is the branch of mechanics dealing with the description of the motion of bodies—the whole body, body segments, or fluids—without reference to the forces producing the motion; whereas the science of *kinetics* studies these forces and other variables related to the motion of a body segment, such as acceleration, motion, or rate of change.

Biomechanical modeling is concerned with developing quantitative techniques to model and study the forces and moments associated with common manual tasks, such as lifting, on the human body. These developments started in the nineteenth century. However, the digital and computer age created excellent opportunities for great advancements in that field.

Anthropometry, as described earlier, is concerned with the collection, tabulation, and statistical analysis of different body-segment measures of populations. New noninvasive techniques are now being adapted to accurately capture and log these measures for a variety of populations.

Bioinstrumentation is the development of new analysis instruments, such as electromyography (EMG), which accurately collects data such as the forces produced by muscles during physical activities; force transducers and force plates; a variety of goniometers to measure angles of motion; and other meters and analysis instruments to measure a variety of forces and a variety of body-segment angles as a result of motion during physical activities (Chaffin, Andersson, and Martin. 1999).

Motion classification and *time prediction systems* and *mechanical work capacities* are important for matching the worker to the task. For example, NASA conducts aggressive studies on the cardiopulmonary functions of astronauts before, during, and after returning from space missions (Sawin et al. 2002). Similarly, some manual lifting tasks and extensive physical activity tasks require extensive screening to match the worker's work capacity to the demands of specific tasks. Again, the main objective is to increase worker productivity while reducing injuries associated with manual work.

Biomechanical Modeling

According to Chaffin, Andersson, and Martin (1999), biomechanical models have been used for various reasons. These include:

- enhancing the academic knowledge to obtain further insight as to how components of a system function and are coordinated to achieve desired outcomes
- providing the medical community with better understanding of how to estimate forces acting on different component body structures used for the prediction of the maximum allowable magnitude for a load held in various postures
- providing engineers with the necessary data to design tools, seats, and workplaces to provide workers with the least stressful configurations
- providing managers with early data to help consider a variety of alternative job conditions, work methods, and personnel stereotypes
- predicting potentially hazardous loading conditions on certain musculoskeletal tissues.

As a result, a number of biomechanical models have been developed and used over the past fifty years. These include planar static biomechanical models, three-dimensional modeling of static strength, dynamic biomechanical modeling, and special-purpose biomechanical models for occupational tasks.

Planar static biomechanical models include single-body-segment static models, such as those used to measure the forces acting on the one hand as a result of lifting an object. The single body segment in this case would be the forearm (see Figure 5).

Planar static biomechanical models also include two-body-segment static models, such as those used to measure the forces acting on the shoulder as a result of a load being applied at the hand. The two body segments in this case would be the forearm (hand to elbow) and the upper arm (elbow to shoulder).

The static planar model of nonparallel forces, such as those produced by an external force acting on the hand, is included and is at angle (α) to the vertical plane.

Planar static analysis of internal forces, such as those produced by the muscles at a distance of the joint to produce a state of static equilibrium, is also included. In this model, the forces in the single-body-segment static model are used to find the internal forces (F_M) exerted by the muscle at a distance (m) from the point of fulcrum of the elbow to maintain a state of static equilibrium (see Figure 5).

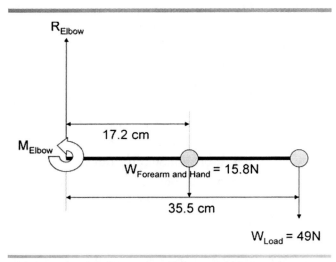

FIGURE 5. Single-body-segment static model of the forearm (Adapted from Chaffin, Andersson, and Martin 1999)

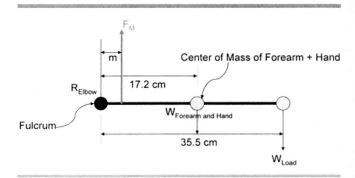

FIGURE 6. Planar static analysis of internal forces (Adapted from Chaffin, Andersson, and Martin 1999)

QUESTION FOR STUDY

Find the reactive forces and moment acting on the elbow as a result of a 5-kg weight lifted by the hand of an average-sized male, as shown in Figure 5. According to Newton's second law,

$$F = ma \quad (1)$$

where

F = Force due to the load, or weight of the load (W_{Load})

m = Mass of the load (5 kg)

a = Acceleration due to gravity (9.8 m/sec²)

Therefore, W_{Load} = 5 kg × 9.8 m/sec² = 49 N acting downward.

From anthropometric tables, the weight of the forearm and hand of an average-sized male is 15.8 N downward.

Finding R_{Elbow}:

Since this assembly is in static equilibrium, then the Σ_{Force} in the y-axis direction is 0. Hence,

$$\Sigma_{Force} = 0 \quad (2)$$

where

Σ_{Force} = The summation of all forces acting in the direction of the y-axis.

Therefore, $R_{Elbow} - (W_{Load} + W_{Forearm\ and\ Hand}) = 0$.

Therefore, $R_{Elbow} = W_{Load} + W_{Forearm\ and\ Hand}$ = 49 N + 15.8 N = 64.8 N acting upward (in the opposing direction to the other two forces).

Finding M_{Elbow}:

Since this assembly is in static equilibrium, then the Σ_{Moment} about the elbow is 0. Hence,

$$\Sigma M_{Elbow} = 0 \quad (3)$$

where

Σ_{Moment} = The summation of all moments acting in the clockwise direction.

Then, $M_{Elbow} - [(W_{Load} \times 35.5\ cm) + (W_{Forearm\ and\ Hand} \times 17.2\ cm)] = 0$.

Therefore, $M_{Elbow} = (W_{Load} \times 0.355\ m) + (W_{Forearm\ and\ Hand} \times 0.172\ m) = (49\ N \times 0.355\ m) + (15.8\ N \times 0.172\ m)$ = 20.11 Nm counterclockwise (in the opposite rotation of the other two moments).

For the previous example, using a load of 5 kg, to find the forces exerted by the muscle (F_M) to maintain a state of static equilibrium, the Σ_{Moment} about the fulcrum of the elbow should be 0 (see Figure 6). Then

$$W_{Load} \times 35.5\ cm + W_{Forearm\ and\ Hand} \times 17.2\ cm = F_M \times m \quad (4)$$

where

W_{Load} = 49 N

$W_{Forearm\ and\ Hand}$ = 15.8 N

m = 5 cm (or 0.05 m)

Therefore, F_M = (49 N × 0.355 m + 15.8 N × 0.172 m) ÷ 0.05 m = 402.25 N positive (upward).

The multiple-link coplanar static model is very similar to the above-mentioned two-body-segment static models, with additional multiple links. In this case, the body forces and moments are analyzed through the analysis of six links connecting six segments together (see Figure 7). These segments are: (1) the forearm, (2) the upper arm, (3) the upper torso, (4) the upper leg, (5) the lower leg, and (6) the foot. External loads are then added into the calculations based on the different angles and postures of the body. The links would naturally be: (1) the elbow, connecting the forearm to the upper arm; (2) the shoulder, connecting the forearm to the upper torso; (3) the hip, connecting the upper torso to the upper leg; (4) the knee, connecting the upper leg to the lower leg; (5) the ankle,

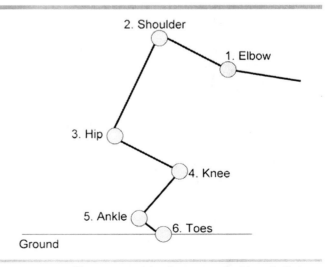

FIGURE 7. Six-segment body connected by six links (Adapted from Chaffin, Andersson, and Martin 1999)

connecting the lower leg to the knee; and (6) the toes, connecting the foot to the ground (see Figure 7).

Three-Dimensional Modeling of Static Strength

While two-dimensional modeling seemed simple and effective in solving and analyzing many occupational biomechanical settings, it proved to be inadequate where a worker has motions in the three axes x, y, and z. Examples include a worker pushing a cart with one hand while stabilizing the motion with the other. In cases such as these, the formulas may include the need to solve for six variables. These are the total forces (Σ_F) acting in the direction of each axis, and the total moment (Σ_M) around each axis. According to Chaffin, Andersson, and Martin (1999), Garg and Chaffin (1975) developed a linkage system model (see Figure 8) to represent specific anthropometric data, body postures, and loads of interest.

Outputs from this model include reactive forces and moments at each of the joints of the linkage. This model also computes the forces required by the muscles to perform a particular task, and thus determine the work load and physical strength required of workers performing the task. These complex vector algebraic calculations are now being performed more accurately and much faster by using computer systems, software, and programs such as the 3-D Static Strength Predicting Program (3D SSPP) developed by the University of Michigan's Department of Industrial Engineering (Chaffin, Andersson, and Martin 1999).

Dynamic Biomechanical Modeling

Dynamic biomechanical modeling presents different complexities from those presented by static modeling, such as kinematic measurements for the direction, velocity, and acceleration of body segments, as well as kinetic measurements for the forces and moments acting on these body segments.

Single-segment dynamic biomechanical models: Similar to single-segment static models, dynamic models present the same set of variables in addition to those introduced to the motion of a body segment. These new variables include inertia, radius of gyration, and centrifugal forces; hence, centripetal forces, velocity and acceleration, and the effects of that rotation on

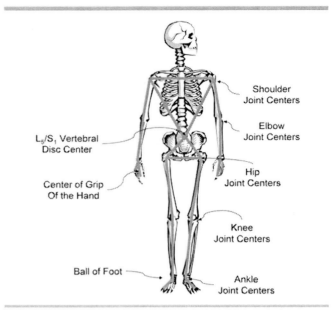

FIGURE 8. Linkage representation for whole-body biomedical modeling (Adapted from Chaffin, Andersson, and Martin 1999)

the joint centers adjoining that body segment. An example of the utilization of this model is modeling the forces and moments on the forearm resulting from lifting an object.

Multiple-segment biodynamic model of load lifting: Similar to multi-segment static modeling, biodynamic modeling calculates the forces and reactive moments resulting from a single segment's motion, then works its way backward to calculate the resulting forces on the adjoining segments and joint centers, with the addition of other kinematic and kinetic considerations similar to those obtained in single-segment dynamic biomechanical calculations. An example of the utilization of this model is modeling the forces, moments, velocities, acceleration, inertia, and other variables from the hand to the different joint centers, such as elbow, shoulder, hips, knees, and ankle joints via the different segments, such as the forearm, the upper arm, the upper torso, the upper leg, the lower leg, and the foot, as a result of lifting an object from the floor to a table.

Coplanar biomechanical models of foot slip potential while pushing a cart: In this model, new external variables are introduced. These are the reacting forces of the cart on the worker's hand as a result of the push

Principles of Ergonomics

force, shear forces on the foot due to friction with the floor, and friction forces due to the worker's feet on the floor.

Special-Purpose Biomechanical Models for Occupational Tasks

These models use the previously discussed models to determine the required worker capability for manual tasks that are found to be taxing on the body. Special-purpose models have been developed for the wrist area, the back area, and other postures known to produce heavy muscle exertion on certain muscle groups. These models include modeling muscle strength, biomechanical models of the wrist and the hand, and lower-back biomechanical models.

Modeling muscle strength: The main purpose of modeling muscle strength is to predict human strength, based on age and gender, during a variety of tasks, lifting being one of them. While static muscle strength has been tested and examined, dynamic models that predict muscle strength are still under development.

Static models, as described earlier, state that, to maintain a state of equilibrium at a joint, the moment around that joint produced by the muscle should be equal to or higher than the moment produced by the external force exhibited by the load. Using this concept, a variety of data has been produced to predict the strength required by an individual (based on the average capability of each gender) using a variety of postures (Chaffin, Andersson, and Martin 1999).

For example, according to Chaffin, Andersson, and Martin, based on research done by Schanne and then corrected for population strength by Stobbe, the predicted shoulder flexion mean strength (S_S) in Newton meters (Nm) is given by this formula (see Figure 9):

$$S_S = [227.338 + 0.525\, \alpha_E - 0.296\, \alpha_S][G] \quad (5)$$

where

α_E = Angle of the elbow joint
α_S = Angle of the shoulder joint
G = Gender adjustment = 0.1495 (for females), 0.2848 (for males)

Calculating the shoulder flexion mean strength (S_S) in Nm for a female when the angle of the elbow

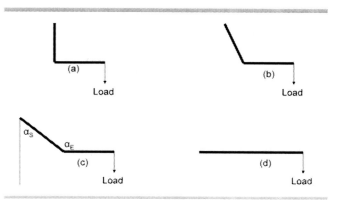

FIGURE 9. Postures: shoulder angles and elbow angles (Adapted from Chaffin, Andersson, and Martin 1999)

joint is 45° and the angle of the shoulder joint is 45°, since,

$$S_S = [227.338 + 0.525\, \alpha_E - 0.296\, \alpha_S][G]$$

then $S_S = [227.338 + 0.525 \times 45° - 0.296 \times 45°][0.1495] = 35.53$ Nm.

Table 1 shows different predicted shoulder mean strengths at different postures. As observed in the table, when substituting the different angles into the formula, the predicted shoulder mean strength for females is relatively constant at an average of around 42.5 Nm. Therefore, with the elbow extended (posture d in Figure 9), one may determine the maximum load capacity of a person by using the values for posture d in Table 1. Calculations show that this load should not exceed 69 N. Similarly, with the elbow close to the body, (posture a in Figure 9), the load should not exceed 155 N.

In addition to the average shoulder strength of females, figures such as these can be calculated for

TABLE 1

Predicted Shoulder Mean Strength at Different Postures

Posture	S_S (Nm)	α_E (degrees)	α_S (degrees)
a	41.050906	90	0
b	42.077971	120	30
c	43.105036	150	60
d	44.132101	180	90

(Adapted from Chaffin, Andersson, and Martin 1999)

other joints in the body, including the elbow, lower back, hip, knee, and ankle for both males and females. These predictions are very important in determining the amount of strength needed to perform a particular task, and the maximum loads that an individual may be able to handle. This may prove to be important for writing job descriptions, worker screening, and for manual-labor job placement to avoid potential injuries.

Biomechanical Models of the Wrist and Hand

Disorders of the wrist and hand have been followed and studied for the past 40 years. Expenses and lost work days as a result of these disorders are known to be costly. The most prevalent of these disorders include carpal tunnel syndrome, ganglionic cysts, and wrist tenosynovitis, which is the main cause of trigger finger (all discussed in a previous section of this chapter).

Biomechanical modeling was developed for the finger to measure the amount of force required to grasp a small object by the hand, to push a button with the finger, or to use the finger in trigger action motions, such as those used by spray painters. Data from finger anatomy, anthropometric data, tendons and moment arms, and distances between centers of rotation were analyzed in terms of loads acting at the tip. Models similar to multisegment systems were developed to measure the loads involved at each knuckle. The force required by the tendon to overcome the load applied at the finger tip can be calculated for a variety of hand sizes (see Table 2), where, F_t = force exerted by the tendon, and F_L = reactive force exerted by the load.

It is important to note that when forces are exerted by the tendon, it stretches by about 1–2 percent of its original length. Given time, the tendon does come back to its original length. According to Chaffin, Andersson, and Martin, this phenomenon was defined by Abrahams as *residual strain*. These exposures may cause ganglionic cysts and other forms of tendonitis (an inflammation of the tendons).

Regarding the wrist, the risk of developing wrist-related CTDs such as carpal tunnel syndrome, and the severity of such syndromes, was found to be related to the angle of flexion/extension, load, gender, wrist thickness, repetitive grip exertions, particularly with deviated wrists (flexed or extended), and the amount of stress the flexor tendons are exposed to (Silverstein, Fine, and Armstrong 1986). Furthermore, according to Rodgers (1987), grip strength was found to be inversely related to the wrist angle of flexion or extension (see Table 3).

Lower-Back Biomechanical Models

Due to its functional importance and the frequency and high cost related to its injury, the back received a lot of attention from most ergonomists, ergonomic researchers, and the government. Furthermore, injuries associated with the back are the most crippling and agonizing to their victims.

In addition to the NIOSH lifting formula described in the following section of this chapter, many other models were developed. Most researchers focused on the lower back because statistics showed this is where the majority of injuries occur to workers performing material-handling tasks. Also, biomechanical analysis showed that this is where most of the load is being handled. Therefore, it was determined that the lumbo-

TABLE 2

Tendon Grip Force (F_t) by Hand Size Based on Object Size

Object Size	Tendon Grip Force (F_t) by Hand Size	
	Small Hand	Large Hand
Grasping a small object	$F_t = 2.8\ F$	$F_t = 3.7\ F$
Pressing down, or grasping large object	$F_t = 3.1\ F$	$F_t = 4.3\ F$

(Adapted from Chaffin, Andersson, and Martin 1999)

TABLE 3

Grip Strength as It Relates to Wrist Deviation

Deviation	Angle of Deviation	Strength
Neutral	0°	100%
Ulnar	45°	75%
Radial	25°	80%
Flexion	45°	75%
Extreme Flexion	65°	45%
Extension	45°	80%

(Adapted from Chaffin, Andersson, and Martin 1999)

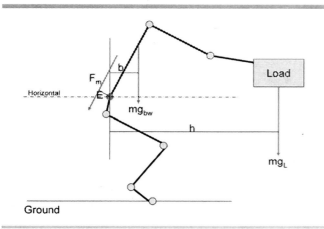

FIGURE 10. Simple lifting model (Adapted from Chaffin, Andersson, and Martin 1999)

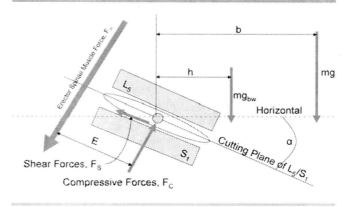

FIGURE 11. Compression and shear forces and angles acting on the L_5/S_1 disk (Adapted from Chaffin, Andersson, and Martin 1999)

sacral region of the back (L_5/S_1) should be used as the basis for setting limits for lifting and carrying tasks to avoid fatigue of the erector spinae muscle group, according to Chaffin, Andersson, and Martin (1999), based on research done by Tichauer.

Other research studies done by Krusen et al. and Armstrong and Smith et al. showed that 85 to 95 percent of all disk herniations occur equally at either L_4/L_5 or L_5/S_1 (Chaffin, Andersson, and Martin 1999).

Morris, Lucas, and Bresler determined there are two internal forces that act to overcome the external load moment. These are: (1) the force exerted by the erector spinae muscles located about 5 cm from the spine (this distance was later revised by other researchers to accommodate different anthropometric data), and (2) the internal abdominal pressure acting at the front of the spine (Chaffin, Andersson, and Martin 1999).

These findings were confirmed by other researchers as well. Chaffin (cited in Chaffin, Andersson, and Martin 1999) adopted the simple cantilever lower-back model of lifting as explained in Figure 8 where there is an imaginary link between the hip joint and the sacrum, and another between the sacrum and the shoulder, to represent the upper torso. Moments around the L_5/S_1 disk were calculated to obtain an idea of the resulting forces by the erector spinae muscle, and hence, compressive and shear forces acting on the L_5/S_1 disk. Figure 10 illustrates how the relationships are governed.

According to Figure 10, for equilibrium, there are no moments acting around the L_5/S_1 disk:

$$\Sigma M_{L_5/S_1} = b(mg_{bw}) + h(mg_L) - E(F_m) = 0 \qquad (6)$$

where

mg_{bw} = Upper body weight and mg_L = Load

F_m = Force acted by the erector spinae muscle group to maintain the body in equilibrium

b = Distance of the vertical component of the force representing the mg_{bw} from the L_5/S_1 disk

h = Distance of the vertical component of the force representing the load from the L_5/S_1 disk

E = Shortest distance of force acted by the erector spinae muscle group and the L_5/S_1 disk

Consider the following data: mg_{bw} = 350 N (upper body weight above L_5/S_1 disk) (Chaffin, Andersson, and Martin 1999):

mg_L = 450 N

F_m = Required

b = 20 cm

h = 30 cm

E = 6.5 cm

Hence, F_m = –3154 N (acting downward, or, according to Figure 10, counterclockwise).

To find the compressive forces and shear forces that the L_5/S_1 disk is exposed to (see Figure 11), one must first determine the angles at which all these forces are acting. Let α be the cutting plane of the L_5/S_1 disk and the horizontal.

According to Andersson (Chaffin, Andersson, and Martin 1999), that angle is equal to

$$40° + \beta$$

where

$\beta = -17.5 - 0.12T + 0.12e^{-2TK} + 0.5e^{-2T^2} - 0.75e^{-3K^2}$
T = Torso angle
K = Knee angle

Therefore, for the above example, if $T = 60°$ and $K = 120°$, then $\beta = 15°$ and $\alpha = 55°$.

For compressive forces:

$$\Sigma F_{Comp} = 0 \qquad (7)$$

$\Sigma F_{Comp} = \cos \alpha (mg)_{bw} + \cos \alpha (mg)_L + F_m - F_C = 0$
$F_C = \cos \alpha (mg)_{bw} + \cos \alpha (mg)_L + F_m$
$F_C = \cos 55(350) + \cos 55(450) + 3154 = 3612$ N (downward).

For shear forces:

$$\Sigma F_{Shear} = 0 \qquad (8)$$

$\Sigma F_{Shear} = \sin \alpha (mg)_{bw} + \sin \alpha (mg)_L - F_S = 0$
$F_S = \sin \alpha (mg)_{bw} + \sin \alpha (mg)_L$
$F_C = \sin 55(350) + \sin 55(450) = 656$ N (forward).

Note that, for the above example, the obtained values fall within the limits of a healthy individual. However, it is noteworthy to mention that these values change according to an individual's age, health, race, gender, and several other factors.

In addition to the previous back models, other models were also developed, including the *finite-element model*, which treats the disk and surrounding structures as a mesh of tensile elements, and calculates the stresses on each element as a result of lifting. According to Chaffin, Andersson, and Martin, other models included 3-D models to account for all other muscle groups in the cutting plane of the lumbar section. These included a ten-muscle 3-D model by Schultz and Anderson; a muscle geometry model where muscles were treated as pointwise connections from origin to insertion by Nussbaum and Chaffin; a vectors model where muscles were treated as 3-D vectors with directions, lengths, and velocities by Marras and Granata; and others as well. Other models involved dynamic lifting models of the back, including dynamic lifting, and dynamic push/pull models.

THE NATIONAL INSTITUTE OF OCCUPATIONAL SAFETY & HEALTH (NIOSH) LIFTING FORMULAS

History and Development of the NIOSH Lifting Equation

The NIOSH lifting formula came about as a natural result of the fast developments in back modeling and analysis. It became a prime example, and one of the first special-purpose biomechanical models, for occupational tasks since manual lifting became a major concern and one of the biggest expenses incurred by employers, workers' compensation agencies, or anyone conducting business in this country in general.

The NIOSH lifting formula was initially published in 1981. However, new research showed that other components such as asymmetrical lifting, quality of hand-container couplings, work durations, and frequency of lifting played important roles in the task of lifting. Hence, a new NIOSH lifting equation was developed in 1991. The main purpose of the NIOSH efforts was to assist safety and health practitioners and ergonomists in assessing lifting tasks in an attempt to reduce material-handling-related back stresses and to stimulate further research and debate on the topic of lower-back pain.

Uses and Limitations of the NIOSH Formulas

NIOSH considered the research results of compression and shear strength of lower-back experiments conducted on cadavers by Genaidy et al. (Chaffin, Andersson, and Martin 1999). As a result, the current NIOSH lifting equation was built for a limit of 3400 N compressive force. This was determined to be the limit that 99 percent of male workers and 75 percent of female workers can withstand for occasional lifting tasks (fewer than 3 lifts/min). The NIOSH lifting equation assumes that:

- nonlifting activities are minimal
- other factors, such as unexpectedly heavy loads, slips, trips, and falls, are not accounted

for, and that environmental temperature and humidity are at comfortable levels (66–79° F and 35–50%, respectively)
- lifting is not conducted by one hand, while seated, kneeling, or constrained, or if the container has a shifting weight, such as those containing liquids
- the foot/floor surface provides at least 0.4–0.5 coefficient of static friction
- lowering or lifting pose the same hazard to the back

According to NIOSH (1997), the NIOSH lifting equation does not apply to one-handed lifting/lowering when
- lifting/lowering for over 8 hours
- lifting/lowering while seated or kneeling
- lifting/lowering in a restricted work space
- lifting/lowering unstable objects
- lifting/lowering while carrying, pushing, or pulling
- lifting/lowering with wheelbarrows or shovels
- lifting/lowering w (more than 30 inch
- lifting/lowering wi least 0.4–0.5 coeffic
- lifting/lowering in environment, as ind

The equations basicall

$$LI = L/RWL \qquad (9)$$

where

$$RWL = LC \times HM \times VM \times DM \times AM \times FM \times CM \qquad (10)$$

The following tables (Tables 4 through 10) provide a list of definitions for the acronyms in the equation, as well as the values to use.

Note that tasks with more activities require the use of the more complicated, multitask analysis.

As an alternative to calculating HM, VM, DM, and AM, the values corresponding to the variables may also be obtained directly from Tables 5 through 8:

TABLE 4

NIOSH Lifting Equation: Acronym Definitions and Values

Acronym	Description	Definition	Value to Use in Equation (English)	Value to Use in Equation (Metric)
L	Load	The actual load being lifted (lbs in English, kg in Metric)	Actual load (in lbs)	Actual load (in kg)
LI	Load Index	L/RWL LI should be ≤ 1	Calculated	Calculated
RWL	Recommended Weight Limit	The highest safe limit that nearly all healthy workers could lift for a substantial period of time (8-hrs) without adverse effects to their health	Calculated (in lbs)	Calculated (in kg)
LC	Load Constant	The highest safe limit that nearly all healthy workers could lift for a substantial period of time (8-hrs) without adverse effects to their health, provided that all other factors and conditions are not affecting the load negatively	51 lbs	23 kg
HM	Horizontal Multiplier	The factor affected by the horizontal distance of the load and the midpoint between the inner ankle bones of the lifter (H)	10/H (H measured in inches)	25/H (H measured in centimeters)
VM	Vertical Multiplier	The factor affected by the vertical distance of travel of the load (V)	1 − (0.0075 |V − 30|) (V is in inches)	1 − (0.003 |V − 75|) (V is in cm)
DM	Distance Multiplier	The factor affected by the horizontal distance of travel of the lifter carrying the load (D)	0.82 + (1.8/D) (D is in inches)	0.82 + (4.5/D) (D is in cm)
AM	Asymmetric Multiplier	The factor affected by the angle of twist of the lifter (A)	1 − (0.0032A) (A is in degrees)	1 − (0.0032A) (A is in degrees)
FM	Frequency Multiplier	The factor affected by the frequency of lifting (F).	From FM table (Table 9)	From FM table (Table 9)
CM	Coupling Multiplier	The factor affected by the coupling provided to the lifter in terms of cutouts, handles, or other means of providing control to the lifter.	From CM table (Table 10)	From CM table (Table 10)

Source: NIOSH 1997

Ergonomics and Human Factors Engineering

TABLE 6

Vertical Multiplier (VM) Table

V English (in)	V Metric (cm)	VM
0	0	0.78
5	13	0.81
10	25	0.85
15	38	0.89
20	51	0.93
25	64	0.96
30	76	1.00
35	89	0.96
40	102	0.93
45	114	0.89
50	127	0.85
55	140	0.81
60	152	0.78
65	165	0.74
70	178	0.70
>70	178	0.00

Source: NIOSH 1997

Horizontal Multiplier (HM) Table

H English (in)	H Metric (cm)	HM
≤10	≤25	1.00
11	28	0.91
12	30	0.83
13	33	0.77
14	36	0.71
15	38	0.67
16	41	0.63
17	43	0.59
18	46	0.56
19	48	0.53
20	51	0.50
21	53	0.48
22	56	0.46
23	58	0.44
24	61	0.42
25	64	0.40
>25	>64	0.00

Source: NIOSH 1997

TABLE 7

Distance Multiplier (DM) Table

D English (in)	D Metric (cm)	DM
≤10	≤25	1.00
15	38	0.94
20	51	0.91
25	64	0.89
30	76	0.88
35	89	0.87
40	102	0.87
45	114	0.86
50	127	0.86
55	140	0.85
60	152	0.85
70	178	0.85
>70	>178	0.00

Source: NIOSH 1997

TABLE 8

Asymmetric Multiplier (AM) Table

A (degrees)	AM
0	1.00
15	0.95
30	0.90
45	0.86
60	0.81
75	0.76
90	0.71
105	0.66
120	0.62
135	0.57
>135	0.00

Source: NIOSH 1997

Principles of Ergonomics

TABLE 9

Frequency Multiplier (FM) Table

Frequency, F (lifts/min)	Work Duration					
	< 1 hr		> 1 hr but < 2hrs		> 2hrs but < 8hrs	
	V < 30	V > 30	V < 30	V > 30	V < 30	V > 30
< 0.2	1.00	1.00	0.95	0.95	0.85	0.85
0.5	0.97	0.97	0.92	0.92	0.81	0.81
1	0.94	0.94	0.88	0.88	0.75	0.75
2	0.91	0.91	0.84	0.84	0.65	0.65
3	0.88	0.88	0.79	0.79	0.55	0.55
4	0.84	0.84	0.72	0.72	0.45	0.45
5	0.80	0.80	0.60	0.60	0.35	0.35
6	0.75	0.75	0.50	0.50	0.27	0.27
7	0.70	0.70	0.42	0.42	0.22	0.22
8	0.60	0.60	0.35	0.35	0.18	0.18
9	0.52	0.52	0.30	0.30	0.00	0.15
10	0.45	0.45	0.26	0.26	0.00	0.13
11	0.41	0.41	0.00	0.23	0.00	0.00
12	0.37	0.37	0.00	0.21	0.00	0.00
13	0.00	0.34	0.00	0.00	0.00	0.00
14	0.00	0.31	0.00	0.00	0.00	0.00
15	0.00	0.28	0.00	0.00	0.00	0.00
> 15	0.00	0.00	0.00	0.00	0.00	0.00

Source: NIOSH 1997

TABLE 10

Coupling Factors (CM Table)

Coupling Type	Coupling Multiplier	
	V < 75cm (30")	V > 75cm (30")
Good: optimal handles	1.00	1.00
Fair: sub-optimal handles	0.95	1.00
Poor: no handles (e.g., bags)	0.90	0.90

Source: NIOSH 1997

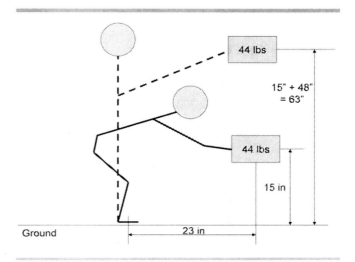

FIGURE 12. Lift "from–to" to illustrate the NIOSH lifting formula

QUESTION FOR STUDY

Given the following, refer to Figure 12 and apply the NIOSH lifting formula.

- Load = 44 lbs
- Hand location away from point connecting ankle bones at start of lift = 23" horizontally, and 15" vertically
- Vertical distance of travel = 48"
- Angle of twist = 0°
- Duration = < 1
- Frequency of lift = < 0.2 lifts/min
- Coupling = Fair

Calculations:

It is important to calculate the *RWL* at both the origin and the destination of the lift to determine the most stressful situation.

Since LI = L/RWL

where
$$RWL = LC \times HM \times VM \times DM \times AM \times FM \times CM$$
and LC = 51 lbs

At the origin:
$HM = 10/H = 10/23 = 0.43$
$VM = 1 - (0.0075 \times V - 30) = 1 - (0.0075 \,|\, 15 - 30 \,|)$
     $= 0.89$
$AM = 1$
$DM = 0.82 + (1.8/D) = 0.82 + (1.8/48) = 0.86$
$FM = 1$ (from Table 9)
$CM = 0.95$ (from Table 10)

Therefore, $RWL = 51 \times 0.43 \times 0.89 \times 0.86 \times 1 \times 1 \times 0.95 = 15.95$ lbs.

$LI = L/RWL = 44/15.95 =$ **2.76**

At the destination:
$HM = 10/H = 10/23 = 0.43$

$VM = 1 - (0.0075 \,|V - 30|) = 1 - (0.0075\,|63 - 30|)$
$= 0.75$
$AM = 1$
$DM = 0.82 + (1.8/D) = 0.82 + (1.8/48) = 0.86$
$FM = 1$ (from Table 9)
$CM = 0.95$ (from Table 10)

Therefore, $RWL = 51 \times 0.43 \times 0.75 \times 0.86 \times 1 \times 1 \times 0.95 = 13.44$ lbs.

$LI = L/RWL = 44/13.44 = \mathbf{3.27}$

Priorities should be made toward the worst of the two situations. In this example, that is the destination. The situation at the origin is not much better, either.

Therefore, to improve this situation, one must consider increasing the smallest multipliers that are lowering the value of RWL and, in turn, lowering the value of LI.

From the example, since the following multipliers were significantly low, recommendations should be geared toward increasing their values: $HM = 0.43$ (at both the origin and the destination), $VM = 0.89$ and 0.75 (at the origin and destination, respectively), $DM = 0.86$ (at both the origin and the destination), and $CM = 0.95$ (at both the origin and the destination, respectively). Therefore, to increase these values, the following suggestions could be applied:

- Bring the object closer to the body of the worker to reduce H and therefore increase HM
- Lower the destination of the lift to increase VM and DM
- Provide for better coupling.

Other Tools for Measuring Ergonomic Stresses

In addition to the NIOSH lifting formula, other tools have also been developed for analyzing and quantifying exposures to ergonomic stresses. These tools include rapid upper limb assessment (RULA), rapid entire body assessment (REBA), the strain index, THERBLIGS, and other psychophysical measurements, such as the Snook Tables, for evaluating manual lifting.

The process of capturing data to use with any of the previously mentioned tools has also been technologically enhanced by the use of gesturing tools, such as apps for iPads, iPhones, iTouches, or other similar tools that facilitate data entry for analysis.

In addition to the NIOSH lifting formula, RULA has been recognized by OSHA for ergonomic analysis of risk factors that have been shown to contribute to ergonomic-related injuries. Originally designed to assess employees who may be exposed to the risk factors known to contribute to upper limb disorders, RULA also provides a method for screening large numbers of employees quickly, and is considered a scoring system for determining the loading experienced by a worker.

According to RULA, the upper body is broken down into two groups, Group A and Group B. Group A includes the upper arms, lower arms, and wrists. Group B includes the neck, trunk, and legs (see Figures 13–19) (McAtamney and Corlett 2004).

In Figure 13: for upper arms, add 1 if the shoulder is raised, add 2 if the upper arm is abducted, and subtract 1 if leaning or supporting the weight of the arm.

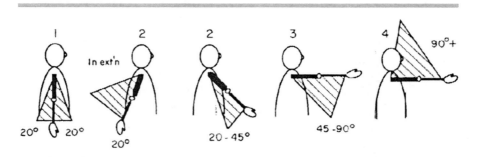

FIGURE 13. Group A: upper arm (*Source:* McAtamney and Corlett 2004)

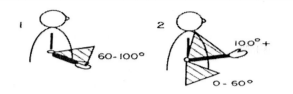

FIGURE 14. Group A: lower arm
(*Source:* McAtamney and Corlett 2004)

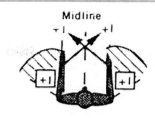

FIGURE 15. Group A: lower arm across midline
(*Source:* McAtamney and Corlett 2004)

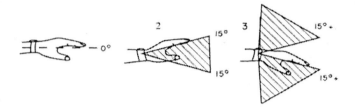

FIGURE 16. Group A: wrists (*Source:* McAtamney and Corlett 2004)

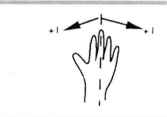

FIGURE 17. Group A: wrist bent away from midline
(*Source:* McAtamney and Corlett 2004)

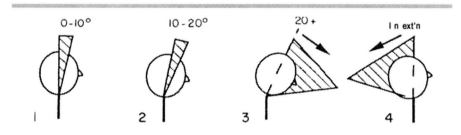

FIGURE 18. Group B: neck (*Source:* McAtamney and Corlett 2004)

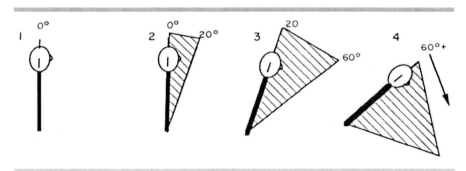

FIGURE 19. Group B: trunk (*Source:* McAtamney and Corlett 2004)

In Figure 14: for the lower arm, add 1 if working across the midline of the body or out to the side.

In Figure 16: add 1 if mainly in the mid-range of the twist, and 2 at or near the end of twisting range. Add 1 if the wrist is bent away from the midline.

In Figure 18: add 1 if the neck is twisted, and add 1 if the neck is bent to the side.

In Figure 19: add 1 if the trunk is twisted, and add 1 if the trunk is bent to the side.

TABLE 11

RULA Score: Group A

Upper Arm	Lower Arm	Wrist Posture Score							
		1		2		3		4	
		Wrist 1	Twist 2	Wrist 1	Twist 2	Wrist 1	Twist 2	Wrist 1	Twist 2
1	1	1	2	2	2	2	3	3	3
	2	2	2	2	2	3	3	3	3
	3	2	3	3	3	3	3	4	4
2	1	2	3	3	3	3	4	4	4
	2	3	3	3	3	3	4	4	4
	3	3	4	4	4	4	4	5	5
3	1	3	3	4	4	4	4	5	5
	2	3	4	4	4	4	4	5	5
	3	4	4	4	4	4	5	5	5
4	1	4	4	4	4	4	5	5	5
	2	4	4	4	4	4	5	5	5
	3	4	4	4	5	5	5	6	6
5	1	5	5	5	5	5	6	6	7
	2	5	6	6	6	6	7	7	7
	3	6	6	6	7	7	7	7	8
6	1	7	7	7	7	7	8	8	9
	2	8	8	8	8	8	9	9	9
	3	9	9	9	9	9	9	9	9

(*Source:* McAtamney and Corlett 2004)

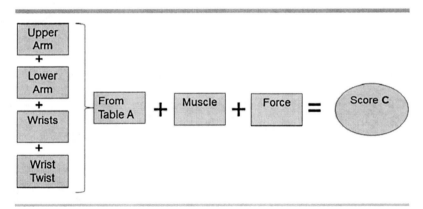

FIGURE 20. RULA Group A scoring (*Source:* McAtamney and Corlett 2004)

Group B-Legs: add 1 if the legs and feet are well supported and in an evenly balanced posture, add 2 if they are not.

Muscle Score: add 1 if the posture is mainly static (e.g., held for longer than 1 minute) or if the motion is repeated more than 4 times/minute.

Forces Score:
- 0: no resistance or less than 2 kg intermittent load or force
- 1: 2–10 kg intermittent load or force
- 2: 2–10 kg static or repeated load
- 3: 10 kg or more static or repeated loads or forces, or shock or forces with rapid buildup

This ergonomic evaluation tool results in a score ranging between one and seven, as seen in Table 13. These scores are used to prioritize jobs that are at a significantly high level of ergonomic exposure. The higher the score, the more "risky" the exposure.

TABLE 12

RULA Score: Group B

Neck Posture Score	Trunk Posture Score											
	1		2		3		4		5		6	
	Legs		Legs		Legs		Legs		Legs		Legs	
	1	2	1	2	1	2	1	2	1	2	1	2
1	1	3	2	3	3	4	5	5	6	6	7	7
2	2	3	2	3	4	5	5	5	6	7	7	7
3	3	3	3	4	4	5	5	6	6	7	7	7
4	5	5	5	6	6	7	7	7	7	7	8	8
5	7	7	7	7	7	8	8	8	8	8	8	8
6	8	8	8	8	8	8	8	9	9	9	9	9

(*Source:* McAtamney and Corlett 2004)

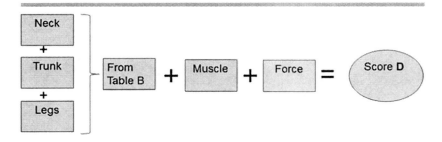

FIGURE 21. RULA Group B scoring (*Source:* McAtamney and Corlett 2004)

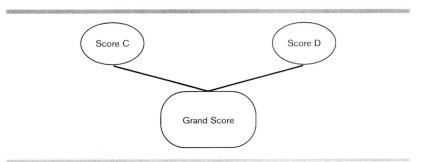

FIGURE 22. RULA grand score (*Source:* McAtamney and Corlett 2004)

TABLE 13

RULA Grand Score

		Score D (Neck, Trunk, Leg)						
		1	2	3	4	5	6	7
Score D (Upper Limb)	1	1	2	3	3	4	5	5
	2	2	2	3	4	4	5	5
	3	3	3	3	4	4	5	6
	4	3	3	3	4	5	6	6
	5	4	4	4	5	6	7	7
	6	4	4	5	6	6	7	7
	7	5	5	6	6	7	7	7
	8	5	5	6	7	7	7	7

(*Source:* McAtamney and Corlett 2004)

REFERENCES

American Academy of Orthopedic Surgeons (AAOS). 2005. *Your Orthopaedic Connection.* orthoinfo.aaos.org/main.cfm

British Broadcasting Corporation. 2007. *Historic Figures: Imhotep.* www.bbc.co.uk/history/historic_figures/imhotep.shtml

Chaffin, Don B., Gunnar B. J. Andersson, and Bernard J. Martin. 1999. *Occupational Biomechanics.* 3d ed. New York: John Wiley & Sons.

Eastman Kodak Company. 1983. *Kodak's Ergonomic Design for People at Work.* vol 1. New York: Van Nostrand and Reinhold.

Eastman Kodak Company. 1986. *Kodak's Ergonomic Design for People at Work.* vol 2. New York: Wiley & Sons, Inc.

Eaton, C. 1997. *E-hand.com—The Electronic Textbook of Hand Surgery.* www.e-hand.com

International Ergonomics Association (IEA). 2000. *What Is Ergonomics?* (retrieved September 15, 2011) www.iea,cc/01_what/What20%Ergonomics.htlm

McAtamney, L., and E. N. Corlett. 2004. "Rapid Upper Arm Assessment (RULA)" in Stanton, N. et al., eds., *Handbook of Human Factors and Ergonomics.* Boca Raton, FL: CRC Press.

Medical Multimedia Group, LLC. 2001. eOrthopod.com. www.eorthopod.com/public

Meyer, Donald W. 2001. "Correction of Lordotic/Kyphotic S-Curves Without Extension Traction." *American Journal of Clinical Chiropractic,* April 2001. www.idealspine.com/pages/AJCC/AJCC_new/April2001/traction.htm

National Institute for Occupational Safety and Health (NIOSH). 1997. Publication 97-141: *Musculoskeletal Disorders and Workplace Factors—A Critical Review of Epidemiologic Evidence for Work-Related Musculoskeletal Disorders of the Neck, Upper Extremity, and Low Back.* www.cdc.gov/niosh/docs/97-141/

Netter, Frank H. 1997. *Atlas of Human Anatomy.* 2d ed. Summit, N.J.: Novartis Pharmaceuticals Corp.

Patel, M. R. 1997a. *Cubital Tunnel Syndrome.* www.handsurgeon.com/cubital.html

Patel, M. R. 1997b. *Thoracic Outlet Syndrome.* www.handsurgeon.com/thoracic_outlet

Patel, M. R. 1997c. *Trigger Finger.* www.handsurgeon.com/trigger.html

Putz-Anderson, Vern, ed. 1988. *Cumulative Trauma Disorders: A Manual for Musculoskeletal Diseases of the Upper Limbs.* New York: Taylor & Francis.

Ramazzini, Bernardino. 1713. *De Morbis Artificum* or *The Diseases of Workers.* Translated from the Latin by Wilmer Cove Wright (1964). New York: Hafner Publishing Company.

Rodgers, S. H. 1987. "Recovery Time Needs for Repetitive Work." *Seminars in Occupational Medicine* 2(1):19–24.

Sanders, Mark S., and Ernest J. McCormick. 1993. *Human Factors in Engineering and Design.* 7th ed. New York: McGraw Hill.

Sawin, Charles F., Arnauld E. Nicogossian, A. Paul Schachter, et al. 2002. *Pulmonary Function Evaluation During and Following Skylab Space Flights.* www.lsda.jsc.nasa.gov/books/skylab/Ch37.htm

Sechrest, R. 1996a. *CTD: Carpal Tunnel Syndrome.* www.healthpages.org/AHP/LIBRARY/HLTHTOP/CTD/cts.htm

Sechrest, R. 1996b. *Thoracic Outlet Syndrome.* www.eorthopod.com/eorthopodV2/index.php/fuseaction/topics.detail/ID/79791a8f7dd9f446b38653cbeab9a955/TopicID/44a14668f9e0b1eafc5748bb059b5c60/area/6

Silverstein, B. A., L. J. Fine, and T. J. Armstrong. 1986. "Hand Wrist Cumulative Trauma Disorders in Industry." *British Journal of Industrial Medicine* 43:779–784.

Spinal Motion, Inc. 2005. *Kineflex.* www.SpinalMotion.com

Spine-Health.com. 2005. *Spinal Anatomy and Back Pain.* www.spine-health.com/topics/anat/a01.html

Wickens, Christopher D., Sallie E. Gordon, and Yili Liu. 1998. *An Introduction to Human Factors Engineering.* New York: Longman.

ADDITIONAL RESOURCES

Abrahams, M. 1967. "Mechanical Behavior of Tendon in Vitro: A Preliminary Report." *Med. Bio. Eng.* 5:433.

Anderson, C. K., and D. B. Chaffin. 1986. "A Biomechanical Evaluation of Five Lifting Techniques." *Applied Ergonomics* 17(1):2–8.

Armstrong, J. R. 1965. *Lumbar Disc Lesions.* Baltimore, MD: Williams and Wilkins.

Board of Certified Professional Ergonomists. BCPE website. www.bcpe.org

Chaffin, D. B. 1987. "The Role of Biomechanics in Preventing Occupational Injury." Special Section, Conference on Injury in America. *Public Health Rep.* 102(6):599–602.

Chehalem Physical Therapy, Inc. 2000. *A Patient's Guide to Rehabilitation for Cubital Tunnel Syndrome.* Newberg, OR: Chehalem Physical Therapy, Inc.

Ergoweb, Inc. 2005. *History of Ergonomics.* www.ergoweb.com/resources/faq/history.cfm

Frankel, V. H., and M. Nordin. 1980. *Basic Biomechanics of the Skeletal System.* Philadelphia: Lea and Febiger.

Garg, A., D. B. Chaffin, and A. Freivalds. "A Biomechanical Computerized Simulation of Human Strength." AIIE Trans. March 1–15, 1975.

Genaidy, A. M., S. M. Waly, T. M. Khalil, and J. Hidalgo. 1993. "Spinal Compression Tolerance Limits for the

Design of Manual Manufacturing Handling Operations in the Workplace." *Ergonomics* 36(4):415–434.

Goldberg, Robert L. 2001. *The Medical Management of Upper Extremity Injury in the Occupational Setting.* www.neurometrix.com/papers_monos%20pdf/GoldbergMonograph.pdf

Human Factors and Ergonomics Society. HFES Web site. www.hfes.org

Kroemer, K. H. E. 1989. "Cumulative Trauma Disorders: Their Recognition and Ergonomics Measures to Avoid Them." *Journal of Applied Ergonomics* 20(4):274–280.

Kroemer, K. H. E., and E. Grandjean. 1999. *Fitting the Task to the Human.* 5th ed. Philadelphia, PA: Taylor & Francis.

Krusen, F., C. M. Ellwood, and F. J. Kottle. 1965. *Handbook of Physical Medicine and Rehabilitation.* Philadelphia: Sanders.

Maisel, M., and L. Smart. 1997. *Women in Science: A Selection of 16 Significant Contributors.* www.sdsc.edu/ScienceWomen/GWIS.pdf

Marras, W. S., and K. P. Granata. 1995. "A Biomechanical Assessment and Model of Axial Twisting in the Thoracolumbar Spine." *Spine* 20(3):1440–1451.

Martin, M. P. 2005. "Holistic Ergonomics: A Case Study From Chevron Texaco." *Journal of Professional Safety* (February) pp. 18–25.

McAtamney, L., and E. N. Corlett. 1993. "RULA: A Survey Method for the Investigation of Work-Related Upper Limb Disorders." *Applied Ergonomics* 24:91–99.

Morris, J. M., D. B. Lucas, and B. Bresler. 1961. "Role of the Trunk in Stability of the Spine." *J. Bone Joint Surg.* 43A:327.

Nussbaum, M. A., and D. B. Chaffin. 1996. "Development and Evaluation of a Scalable and Deformable Geometric Model of the Human Torso." *Clinical Biomechanics* 11(1):25–34.

Roth, C. L. 2005. "How to Protect the Aging Work Force." *Occupational Hazards* (January) pp. 38–42.

Schanne, F. A. 1972. "A Three-Dimensional Hand Force Capability Model for the Seated Operator." Unpublished doctoral dissertation. Ann Arbor, MI: University of Michigan.

Schultz, A. B., and B. J. G. Andersson. 1981. "Analysis of Loads on the Lumbar Spine." *Spine* 6(1):76–82.

Schultz, A. B., B. J. G. Andersson, R. Ortengren, K. Haderspeck, and A. Nacherson. 1982. "Loads on the Lumbar Spine." *J. Bone Joint Surg.* 64-A:713–720.

Smith, A., E. M. Deery, and G. L. Hagman. 1944. "Herniations of the Nucleus Pulposus: A Study of 100 Cases Treated by Operation." *J. Bone Joint Surg.* 26:821–833.

Stobbe, T. 1982. "The Development of a Practical Strength Testing Program for Industry." Unpublished doctoral dissertation. Ann Arbor, MI: University of Michigan.

Tichauer, E. R. 1971. "A Pilot Study of the Biomechanics of Lifting in Simulated Industrial Work Situations." *J. Safety Res.* 3(3):98–115.

Waters, Thomas R., Vern Putz-Anderson, and Arun Garg. 1994. *Applications Manual for the Revised NIOSH Lifting Equation* (NIOSH Publication 94-110). Cincinnati, OH: National Institute for Occupational Safety and Health, Division of Biomedical and Behavioral Science.

Waterson, P., and R. Sell. 2007. *Chronology of the Society.* The Ergonomics Society. About the Society. www.ergonomics.org.uk/page.php?s=3&p=15

WORK PHYSIOLOGY

3

Carter J. Kerk and Adam K. Piper

LEARNING OBJECTIVES

- Be able to define key terms.

- Apply introductory anthropometric design principles in work design using appropriate data and allowances.

- Understand the fundamentals of various systems in the body, including the skeletal, skeletal muscular, neuromuscular, respiratory, circulatory, and metabolic systems.

- Estimate energy requirements for walking and lifting.

- Understand the importance of aerobic work design.

- Be able to estimate oxygen consumption or uptake.

- Understand localized muscle fatigue, general physiologic and mental fatigue, and be able to identify possible control measures.

- Apply introductory work design principles to minimize fatigue and enhance performance.

- Understand work schedules and circadian rhythms.

WORK PHYSIOLOGY IS the study of physiological information about humans and how to apply that information in the evaluation and design of work. The term *physiology* is defined as the study of the processes and functions of an organism—in this case, the human organism. The realm of work physiology (including anthropometry for this chapter) includes: body systems (skeletal, skeletal muscular, neuromuscular, respiratory, circulatory, and metabolic), thermal stress, evaluation of cardiovascular capacity, fatigue, and work design. Work physiology and anthropometry can help the practicing safety professional to minimize occupational injuries while providing a safer workplace and improving productivity.

ANTHROPOMETRY
Definition and Use

The term *anthropometry* literally means "the measure of humans." From a practical standpoint, the field of anthropometry is the science of measurement and the art of application that establishes the physical geometry, mass properties, and strength capabilities of the human body (Roebuck 1995). Anthropometric data are fundamental in the fields of work physiology (Åstrand et al. 2003), occupational biomechanics (Chaffin, Andersson, and Martin 2006), and ergonomic/work design (Konz and Johnson 2004). Anthropometric data are used in the evaluation and design of workstations, equipment, tools, clothing, personal protective equipment (PPE), and products, as well as in biomechanical models and bioengineering applications.

Anthropometric Data

It is a fundamental concept of nature that humans come in a variety of sizes and proportions. Because there is a reasonable amount of useful anthropometric data available, it is usually not necessary to collect measurements on a specific workforce. The most common application involves design for a general occupational population. Some selected anthropometric data (body dimensions) are shown in Table 1. These data were collected on seminude subjects in rigid, erect postures; therefore, certain allowances must be applied for most practical uses. For shoe height, add 1 inch; for shoe weight, add 2 pounds; and for clothing weight, add 1 pound (Marras and Kim 1993). People rarely work in rigid, erect postures, so allowance for slumping may be appropriate for standing (subtract 1 inch) and sitting positions (subtract up to 1.5 inches) (Pheasant and Haslegrave 2006). For a comprehensive source of anthropometric data, Pheasant and Haslegrave (2006) and Roebuck (1995) provide extensive information about statistical aspects, data collection methods, gender differences, ethnic differences, and aging trends.

Anthropometric Design Principles

There are three general anthropometric design principles useful in the design of workspaces. Each design principle is described with its advantages and disadvantages.

1. **Design for Average.** With the *design for average* principle, a workspace is designed for the average-sized person—a one-size-fits-all approach. It is commonly used by designers without knowledge of population variability and is generally not recommended. Normally it presents the least-cost method, another reason it is commonly used. For example, a designer might design a standing workstation for light assembly work at the average standing elbow height for the 50th percentile male and female: $[(39.3 + 42.2) \div 2] + 1 - 1 = 40.75$ inches above

TABLE 1

Body Dimensions (Inches) of Seminude U.S. Adult Civilians

	Percentile						Standard Deviation Female	Standard Deviation Male
	5th Female	5th Male	50th Female	50th Male	95th Female	95th Male		
Heights (above floor)								
Height (standing)	60.1	64.8	64.1	69.1	68.4	73.5	2.50	2.63
Eye height	55.7	60.2	59.7	64.3	63.8	68.6	2.46	2.59
Shoulder height	48.9	52.8	52.5	56.8	56.4	60.9	2.28	2.44
Elbow height	36.5	39.2	39.3	42.2	42.3	45.4	1.76	1.89
Knuckle height[1]	26.4	27.6	28.7	30.1	31.1	32.7	1.46	1.61
Sitting Heights (above seat)								
Height (sitting)	31.3	33.6	33.5	36.0	35.8	38.3	1.37	1.40
Eye height	27.0	28.9	29.1	31.2	31.3	33.4	1.31	1.35
Shoulder height	20.0	21.6	21.9	23.5	23.8	25.4	1.13	1.17
Elbow height	6.9	7.2	8.7	9.1	10.4	10.8	1.06	1.07
Thigh height	5.5	5.9	6.3	6.6	7.1	7.5	0.48	0.50
Popliteal height (above floor)	13.8	15.5	15.3	17.1	16.9	18.8	0.93	0.98
Depths								
Forward reach (thumbtip)	26.6	29.1	28.9	31.5	31.4	34.1	1.43	1.54
Buttock-knee distance (sitting)	21.3	22.4	23.2	24.3	25.2	26.3	1.17	1.18
Buttock-popliteal distance (sitting)	17.3	18.0	19.0	19.7	20.8	21.5	1.05	1.05
Weight (lb)[2]			139.2	182.3				

[1]From Abraham, Johnson, and Najjar (1979)
[2]From Marras and Kim (1993)

(Unless otherwise noted, adapted from Kroemer, Kroemer, and Kroemer-Elbert 2001, p. 27)

floor level. (Remember to add 1 inch for shoe sole allowance and to subtract 1 inch for standing slump.)

The problem with the one-size-fits-all approach is that it fails to accommodate people at both ends of the population distribution, specifically the shortest females and the tallest males. The shortest females forced to work at this assembly workstation will find the surface too high and may develop shoulder discomfort from excessive shoulder flexion. The tallest males will find the surface too low and may develop low back or neck discomfort from extensive flexion. If these discomforts lead to injuries and workers' compensation claims, then this will not be the least-cost method from a systems viewpoint.

2. **Design for Extreme.** The *design for extreme* principle is very useful in specific circumstances when it makes sense to design a dimension at an extreme end of the population distribution and, because of its function, the entire distribution is accommodated. Here are a few examples:

   - A doorway is designed so that extremely tall males and extremely broad people can fit through it. For example, an interior doorway may be designed to be 36 inches wide and 80 inches tall. Both these dimensions exceed the 99th percentile for height and body breadth. A doorway designed for the 50th percentile person would present problems, especially since there are life safety code and emergency egress concerns.
   - If reach distances are designed for the shortest female to reach, then all will be accommodated. Do not design for fingertip reach, but for thumbtip reach, as this is more functional.
   - This principle should also be applied in general for strength requirements, with some precautions. If the weaker person has the strength capability for the task, then, generally, all will be accommodated. In some cases, this may set the strength requirement lower than practical and perhaps be economically

infeasible, so use reasonable judgment. It is the authors' contention that the people in the weakest tail of the strength distribution will self-select out of jobs with moderate-to-high strength requirements.

3. **Design for Range.** *Designing for the range* normally means designing an adjustable workspace. Returning to the standing workstation for light assembly, an adjustable-height workstation might be designed to accommodate elbow heights ranging from the 5th percentile female to the 95th percentile male, or 36.5 to 45.5 inches. Now the potential for shoulder, neck, and lower back discomfort discussed earlier may have been eliminated. Adjustability is one of the keys to effective ergonomic design. The adjustability function and features will come at a greater initial investment, but the potential increases in productivity, worker comfort, and reduced risk of workers' compensation claims will help to make this a favorable investment. Most companies may hesitate to practice the design for range principle in their workplaces initially because they fail to look at the economic advantage. Also, OSHA is not likely to include designing for range in regulations, so companies that pursue regulatory compliance will ignore this principle. However, it is important to note that this principle makes good economic sense and will likely improve productivity and worker satisfaction.

Practical Application of Anthropometric Data

As stated earlier, anthropometric data are useful in work physiology, occupational biomechanics, and ergonomic/work design applications. Of these, the most common practical use is in the ergonomic design of workspaces and tools. Here are some examples:

- The optimal *power zone* for lifting is approximately between standing knuckle height and elbow height, as close to the body as possible. Always use this zone for strategic lifts and releases of loads, as well as for carrying loads.

But minimize the need to carry loads—use carts, conveyors, and workspace redesign.
- Strive to design work that is lower than shoulder height (preferably elbow height), whether standing or sitting. (Special requirements for vision, dexterity, frequency, and weight must also be considered.)
- The upper border of the viewable portion of computer monitors should be placed at or below eye height, whether standing or sitting (Konz and Johnson 2004).
- Computer input devices (keyboard and mouse) should be slightly below elbow height, whether standing or sitting (Konz and Johnson 2004). It has been the authors' experience that this guideline is widely violated, with monitors placed too high, leading to neck discomfort from prolonged, unnecessary neck extension. Bifocal eyeglass users must also be considered.
- Use split keyboards to promote neutral wrist posture (Konz and Johnson 2004). Learn keyboard shortcuts to minimize excessive mouse use. Use voice commands—speech recognition software is increasingly effective for many users and applications.
- For a seated computer workspace, the lower edge of the desk or table should leave some space for thigh clearance (Konz and Johnson 2004).
- For seating, the height of the chair seat pan should be adjusted so the shoe soles can rest flat on the floor (or on a foot rest), while the thighs are comfortably supported by the length of the seat pan (Konz and Johnson 2004). Use knowledge of the popliteal (rear surface of the knee) height, including the shoe sole allowance from Table 1.
- The chair seat pan should support most of the thigh length (while the lower back is well supported by the seat back), while leaving some popliteal clearance (Konz and Johnson 2004). In other words, the forward portion of the seat pan should not press against the calf muscles or back side of the knees. Refer to the seated buttock-popliteal distance in Table 1.
- For horizontal reach distances, keep controls, tools, and materials within the forward reach (thumbtip) distance. Use the anthropometric principle of designing for the extreme by designing the reach distances for the 5th percentile female, thus accommodating 95 percent of females and virtually 100 percent of males.

Body Systems

In applying work physiology in the evaluation and design of work, it is essential to have fundamental knowledge of several relevant body systems, including the skeletal, skeletal muscular, neuromuscular, respiratory, circulatory, and metabolic systems. A brief introduction to each of these systems is presented to provide an elementary foundation. The practitioner is advised to learn these systems in more detail than is presented in this chapter. Recommendations on authoritative references include *Grant's Atlas of Anatomy* (Dalley and Agur 2004), *The Anatomy Coloring Book* (Kapit and Elson 2002), *The Physiology Coloring Book* (Kapit, Macey, and Meisami 2000), *Engineering Physiology* (Kroemer, Kroemer, and Kroemer-Elbert 1997), *Hollinshead's Functional Anatomy of the Limbs and Back* (Jenkins 2002), and *The Extremities: Muscles and Motor Points* (Warfel 1985).

The Skeletal System and Connective Tissue

A complete discussion of the skeletal system should include all *connective tissue*. Connective tissue includes bone, tendons, ligaments, fascia, and cartilage. Connective tissue provides support for the body and structural integrity of body parts and transmits forces (Chaffin, Andersson, and Martin 2006).

Ligaments, *tendons*, and *fascia* are dense connective tissue. Ligaments connect bone to bone and are quite significant in stabilizing joints. Tendons connect bone to muscle. Fascia covers muscle tissue, some internal organs, and holds a significant role in the makeup of the skin layers. Ligaments and tendons consist of dense, parallel fibers—both collagen (inelastic) and elastic—that are capable of powerful axial loads with a slight ability to stretch. The fibers in fascia are irregularly arranged and can resist loading in many directions about the plane, but not perpendicular to the plane, much like a fish net. There is virtually no vascularization (blood flow) to ligaments, tendons, and fascia; therefore its healing ability is limited (Jenkins 2002).

There are two types of cartilage pertinent to our topic: *hyaline* and *fibrocartilage*. Hyaline cartilage covers the end of long bones at *synovial joints*. Synovial joints are lubricated, highly mobile joints throughout the body, including the knuckle, wrist, elbow, shoulder, hip, knee, and ankle joints. (Think of the white cartilage at the end of a chicken leg bone.) Hyaline cartilage is very thin and smooth, protecting bone wear and enhancing smooth joint mobility with the assistance of synovial fluid, while transmitting significant forces between bones.

The major location of fibrocartilage is found in intervertebral discs between each pair of vertebrae in the spinal column. It provides a cushion between the vertebrae and is extremely resistant to compressive forces (up to a point). Fibrocartilage in the intervertebral discs is not well designed to resist tension, shear, and torsion, a fact that provides some insight into some spinal injuries and ideas for proper work design. There is no vascularization (blood flow) in cartilage, but it does receive its nutrition by diffusion. Thus it heals much more slowly or less completely as compared to vascularized tissue (Jenkins 2002).

The final category of connective tissue is *bone*. Bone consists of collagen fibers in a mineralized (calcium) matrix. Bone is well vascularized (has a good blood supply). Bone functions as a support structure, a site of attachment for skeletal muscle, ligaments, tendons, and joint capsules, a source of calcium, and a significant site of blood cell development for the entire body.

There are five classifications of bones: *long* (clavicle, humerus, radius, ulna, phalanges, femur, tibia, and fibula); *short* (really cube-shaped), carpals (wrist) and tarsals (ankle); *flat* (cranial bones, ribs); *irregular* (scapula, pelvis, vertebrae, sternum); and *sesamoid* (patella) (Jenkins 2002). There are approximately 200 bones in the body. Many of these bones intersect at important joints. Synovial joints were discussed earlier. These joints provide for varying types of mobility, as well as allow for the passage of nerves and blood vessels, and, consequently, are sites for constriction or pressure.

TABLE 2

Selected Joint Motions or Functions

Joint Name	Degrees of Freedom	Motion or Function
Interphalangeal (finger and toe joints)	1	flexion, extension
Metacarpal-phalangeal, Metatarsal-phalangeal (hand to finger and foot to toe)	2	flexion, extension abduction, adduction
Wrist	2	flexion, extension radial & ulnar deviation
Elbow	2	flexion, extension forearm supination & pronation
Shoulder	3	flexion, extension abduction, adduction internal & external rotation
Hip	3	flexion, extension abduction, adduction internal & external rotation
Knee	2	flexion, extension internal & external rotation
Ankle	2	plantar flexion, dorsiflexion inversion, eversion

A summary of some of the major mobility joints of the body are presented in a simplified form in Table 2. A single degree of freedom consists of a single paired motion (for example, flexion/extension represent one degree of freedom and abduction/adduction represent another). *Adduction* means to "bring together." *Abduction* means to "move apart." Moving beyond the simplified presentation of joint movements, we find that the shoulder, for example, also has movements of elevation, depression, protraction, and retraction because of the nature of the scapula floating on the posterior rib cage (Jenkins 2002).

A significant component of the skeletal system is the *vertebral column* or *spine*. It consists of four regions from superior (top) to inferior (bottom): *cervical*, *thoracic*, *lumbar*, and *sacral*. The cervical, thoracic, and lumbar regions provide varying (but somewhat limited) amounts of motion in three degrees of freedom (flexion/extension, side bending, and twisting). The spine provides a pathway and protection for the spinal cord, with pairs of spinal nerves emanating between intervertebral joints.

In addition to serving as a pathway and protector for the spinal cord, the vertebral column functions as a support structure for most of the body and is a site of attachment for a multitude of muscles and ligaments. The *cervical* region (neck) consists of seven vertebrae that support the neck and head [weighing approximately 6–8 pounds (2.7–3.6 kilograms)]. They are the most mobile vertebrae and provide the spinal nerves (brachial plexus), serving the upper extremities (the shoulder, upper arm, elbow, lower arm, wrist, hand, and fingers). The *thoracic* region (trunk) consists of twelve vertebrae (progressively larger, moving inferiorly) that support the thorax, neck, and head, as well as articulation with twelve sets of ribs and twelve sets of spinal nerves. The thoracic region is much less mobile than the cervical region, largely because of articulation with the ribs. The *lumbar* region (lower back) consists of the five largest vertebrae that provide support for the entire upper body and torso. The lumbar region is capable of more mobility than the thoracic region. In part because of this mobility and considerable support demand, the lumbar region is the site of most clinical attention for injuries and surgical repair of herniated discs. When a disc ruptures (herniates) or bulges, there is great danger that it can compress against a spinal nerve, causing numbness, tingling, and sharp pain in parts of the body affected by that spinal nerve (sensory dermatomes). The *sacrum* is the final region of the spinal column and is actually a fused set of vertebrae, although the coccyx (tail bone) does float at the terminal end. The lateral sides of the sacrum form strong joints with the pelvis in the sacroiliac (SI) joint. The disc that sits atop the sacrum and below the fifth lumbar vertebra (the L5/S1 disc) supports the most weight of all the discs and is an important point of biomechanical and clinical interest (Jenkins 2002).

The Skeletal Muscular System

The human body possesses three types of muscle: *skeletal*, *smooth* (found in blood vessels and internal organs), and *cardiac* (found in the heart). The purpose of skeletal muscle is to move or stabilize body segments. There are 232 distinct skeletal muscles associated with the extremities, with at least another 42 distinct skeletal muscles devoted to the back. Skeletal muscles possess the ability to contract. Skeletal muscles are composed primarily of voluntary muscle fibers, but also contain small quantities of connective tissue and considerable blood vessels and nerves. Skeletal muscles are generally long and slender, and traverse a skeletal joint. Each end of a muscle is attached to a bone by one or more tendons. The body of the muscle consists of generally parallel muscle fibers (Jenkins 2002).

When a muscle contracts, we normally think of the muscle as shortening in length, exhibiting strength and providing force for work tasks. However, a contracting muscle may also retain its length (a so-called *isometric exertion*) or even lengthen (a so-called *eccentric contraction*). Take the example of holding a bucket with your elbow at a 90-degree angle. If you lift that bucket by flexing your elbow joint, your bicep muscle performs a shortening contraction, while the opposing tricep muscle performs a lengthening contraction. To lower the bucket by extending your elbow joint, the tricep muscle performs a shortening contraction, while the opposing bicep muscle performs a lengthening contraction (Kroemer, Kroemer, and Kroemer-Elbert 1997).

Each muscle consists of hundreds to thousands of muscle fibers that are controlled in small groups by nerve cells. If you require a mild or controlling exertion, your body has learned to activate only a small group of nerve cells, which recruit a relatively small group of muscle fibers. If you require a strenuous, maximal exertion, your body has learned to activate as many nerve cells as possible, recruiting most, if not all, muscle fibers available in that muscle. In either case, these muscle fibers are activated on an *all-or-none principle* (Jenkins 2002). It is not possible to *partially* activate individual muscle fibers. But by the miraculous ability of the body to coordinate motor nerves, it is possible to smoothly control muscles on a continuum from a light, controlling contraction to an all-out maximal contraction. These amazing motor control programs are learned from infancy through adulthood and are practiced and refined thousands of times. Consider that your body flawlessly executes complex motor control programs while climbing stairs or lifting a box. A neurological injury, such as a stroke, may interrupt or erase such motor programs; fortunately they can, in some cases, be relearned.

From an occupational safety and health viewpoint, it is important to focus on four points with regard to skeletal muscles: (1) avoiding extreme exertions, (2) avoiding overly excessive repetitive motions, (3) avoiding awkward postures, and (4) avoiding localized muscle fatigue.

Extreme Exertions: Overexertion from extreme use of force can produce an acute type of injury that can traumatically damage muscle, the muscle-tendon interface, the tendon-bone interface, and possibly rupture adjacent fibrocartilage (spinal discs). A common injury example is a worker lifting an unknown, extremely heavy load. Another injury example is a worker lifting a load that is expected to be light, but instead is quite heavy.

Excessive Repetition: Overexertion from excessive repetition of joint motion can produce a chronic type of injury in the muscles and tendons. It may take weeks, months, or even years for these types of injuries to develop. In general, a job is considered repetitive if the basic cycle time is less than 30 seconds (Konz and Johnson 2004). The combination of repetition with excessive force may produce a detrimental effect that is worse than the sum of the parts.

Awkward Posture: Overexertion of skeletal muscles in awkward postures should be avoided, primarily because of poor mechanical advantage. In posture extremes, muscles may not be able to produce the forces required by the task, may place extreme stress on tendons, and may put pressure on nerve tissue and blood vessels. Awkward postures are usually at the ends of the range of motion for joints. The combination of awkward posture with excessive repetition and with excessive force may produce a detrimental effect that is synergistically worse than the individual components.

Localized Muscle Fatigue: Prolonged isometric muscle exertions should be avoided. In these cases, the muscle motor units are over-used, and the circulatory system is unable to provide oxygen and nutrients to the muscle cells, or remove carbon dioxide and lactic acid (Chaffin, Andersson, and Martin 2006). For example, using the hand as a clamp or vise for extended periods of time should be avoided. Further discussion of fatigue will follow later in the chapter.

The Neuromuscular System

The nervous system is an important control and regulation system in the human body. It gathers input from various sensors throughout the body, both internal and external. It processes information both in the brain and spinal cord to provide regulation of various body functions and control of motor activities. Our interests in this chapter on work physiology focus primarily on thermal regulation and motor control. Thermal regulation will be addressed later in this chapter. This section will take an introductory look at motor control. There is much more to the nervous system, but it is beyond the scope of this chapter.

Anatomically, there are three major subdivisions of the nervous system: (1) the *central nervous system*, (2) the *peripheral nervous system*, and (3) the *autonomic nervous system*. The *central nervous system* (CNS) includes the brain and spinal cord, and maintains primary controls. There are specific portions of the brain that regulate systems and aspects that are critical to the performance of working tasks, including respiration, cardiac function, digestion, attention, thermoregulation, learning, speech, vision, hearing, memory, emotions, and, most pertinent to our discussion, motor control. The CNS receives information from a multitude of sensors in the peripheral nervous system (Kroemer, Kroemer, and Kroemer-Elbert 1997).

The *peripheral nervous system* (PNS) includes the cranial and spinal nerves, transmitting signals to and from the brain along networks of nerve cells, or neurons. The PNS possesses sensors that respond to light, sound, touch, temperature, chemicals, pressure, and pain. Some of the most important sensors for motor control include *proprioceptors*, which provide information about the degree of stretch in muscles, the amount of tension in muscle tendons, the relative location of body joints, and even information about velocity and acceleration of joints. The *vestibular system*, which regulates equilibrium (or balance), is located in the inner ear and consists of three bony, semicircular canals oriented at

90 degrees to one another. The PNS is a two-way system. Thus, just as there are sensors bringing complex information to the CNS, the PNS is also delivering action signals from the CNS for motor control (Kapit and Elson 2002).

The *autonomic nervous system* (ANS) generates the *fright*, *flight*, or *fight response* and regulates involuntary functions such as cardiac muscle, internal organs, blood vessels, and digestion, all of which are critical to the workplace (Kapit and Elson 2002).

Therefore, motor control is regulated from the CNS through the PNS to the muscular system. Some of this was discussed previously in the section entitled "Skeletal Muscle System." The CNS sends signals to specific motor neurons. Each motor neuron controls a set of muscle fibers that, when called upon, will stimulate those muscle fibers to contract with the help of a chemical neurotransmitter (acetylcholine), following the all-or-none principle discussed previously (Kapit and Elson 2002). This is important to the work physiology topic considering the earlier discussion of localized muscle fatigue and the more general discussion of fatigue that follows later in the chapter.

The Respiratory System

The *respiratory system* moves air to and from the lungs from the atmosphere. Within the lungs, part of the oxygen contained in the air is absorbed into the bloodstream. While delivering oxygen to the bloodstream, the air in the lungs simultaneously collects carbon dioxide, water, and heat from the bloodstream and then exhales them to the atmosphere. The respiratory path begins with the mouth and nose, passes through the throat (pharynx), voice box (larynx), and windpipe (trachea), and then enters the bronchial tree. The bronchial tree divides into 23 steps, ending in microscopic spherical-shaped alveoli. There are hundreds of millions of alveoli, which provide as much as 80 square meters of gas exchange surface to the blood circulatory system in an adult (Kroemer, Kroemer, and Kroemer-Elbert 1997). The physiology of the bronchial tree and alveoli are of particular interest in industrial hygiene in the study of airborne contaminants, such as toxins, particulates, aerosols, gases, vapors, dusts, mists, and fumes (Nims 1999). Inspiration and expiration of air by the lungs is powered primarily by the contraction and relaxation of the thoracic diaphragm muscle and, to a lesser degree, by the intercostal muscles (between the ribs) (Kapit and Elson 2002).

Our primary interest in the respiratory system with regard to work physiology is both the quantity and quality of air moved in and out of the lungs. The quantity of air processed by the lungs is a function of body size, gender, age, conditioning, and work demand. Highly trained, large males have a lung capacity of 7–8 liters, of which about 6 liters is usable (vital capacity). Women have lung volumes about 10 percent smaller than their male peers. Untrained persons have volumes of 60–80 percent of their athletic peers. At rest we breathe 10–20 times per minute, increasing to about 45 times per minute for heavy work or exercise (Kroemer, Kroemer, and Kroemer-Elbert 1997).

The quality of air processed by the lungs begins with the concentration of oxygen in the atmosphere. Normal atmospheric air consists of 20.9 percent oxygen. Concentrations less than 19.5 percent are considered oxygen-deficient and are a concern in confined spaces as well as other industrial environments. Without adequate oxygen, workers can become dizzy, uncoordinated, and pass out. Unless this situation is rectified quickly, brain damage or death can occur. Concentrations of more than 23 percent are considered oxygen-enriched and can pose fire or explosion hazards (Nims 1999).

When at rest or during relatively sedentary activity, the body has little demand for oxygen and the exhaled concentration may be 19–20 percent oxygen. During increasingly physically demanding work, the body demands and takes in more oxygen. During extremely demanding work, the exhaled concentration may be as low as 14 percent oxygen. Athletes and trained persons also have the ability to process more oxygen than their less-trained peers at the same workload. This points to the importance of conditioning for workers, especially for physically demanding jobs, and for work-hardening when workers return from extended vacations or injury.

The Circulatory System

The *circulatory system*, using the heart as a pump and blood as the transport medium, plays an important role in transporting materials throughout the body. The blood carries oxygen, nutrients, carbon dioxide, heat, lactic acid, and water to and from various points about the body. The circulatory and respiratory systems are closely related at the alveolar interface in the lungs. The circulatory and metabolic systems are closely related through internal organs such as the stomach, intestines, and liver. The circulatory and musculoskeletal systems are closely related at the capillary interfaces in muscle cells (Kapit, Macey, and Meisami 2000).

The adult body contains 4–6 liters of blood, depending on gender and size. About two-thirds of this volume resides in the venous blood vessels (returning to the heart) and one-third in the arterial vessels (moving away from the heart). An important component of blood, with regard to work physiology, is *hemoglobin*, the iron-containing protein molecule of the red blood cells. Hemoglobin is a willing receptor for oxygen molecules, carbon dioxide, and carbon monoxide (Kapit and Elson 2002).

Working muscles and body organs place demands on the circulatory system for transport of oxygen, carbon dioxide, nutrients, lactic acid, and heat. The heart is the pump for this system. The output (liters per minute) of the heart pump has two primary input factors: the frequency of contraction [or heart rate (HR)] and the stroke volume. At rest, the heart pulses at 50–70 beats per minutes (bpm). This rate increases to as much as 200 bpm at extreme workload levels. Maximum heart rate can be estimated by the Karvonen Formula (Equation 1), which estimates that you lose about one beat per minute for each year of age after age 20. This formula is only an estimate, and individual values can vary by plus or minus 15 bpm (Burke 1998):

$$\text{Heart Rate Maximum} = 220 - \text{Age} \qquad (1)$$

A suggested estimate for older, fit individuals is shown in Equation 2 (Burke 1998):

$$\text{Heart Rate Maximum} = 205 - 0.5(\text{Age}) \qquad (2)$$

In some cases, the heart rate can be used as an indicator of workload, but this is influenced by age, conditioning, and, to some extent, mental stress. Heart rate is a reasonable indicator of workload for light and moderate work demand. At rest, heart rate is influenced greatly by other factors, such as emotions and mental stress and cannot be relied upon as a physical indicator. The cardiac output at rest is about 5 liters per minute (L/min). At strenuous levels of exercise or work, this output can increase by fivefold to about 25 L/min. There are many factors that affect and control the cardiac output, starting with the workload demand, but also the conditioning and health of the heart, the flow resistance in the arteries and veins, and the capillary resistance in the muscle cells and organs (Kroemer, Kroemer, and Kroemer-Elbert 1997).

The four-chambered heart powers two closed-loop circulation systems. The right atrium receives deoxygenated blood from the muscles and organs through the venous return system. The blood is then pumped through the atrioventricular (AV) tricuspid valve to the right ventricle, which pumps the deoxygenated blood through the pulmonary semilunar valve via the pulmonary arteries to the alveolar interface in the lungs, where the hemoglobin exchanges carbon dioxide for oxygen. The oxygenated blood returns through pulmonary veins to the left atrium. The blood is pumped through the AV bicuspid (mitral) valve to the left ventricle, which in turn pumps the oxygenated blood through the aortic semilunar valve via the aorta and into the arterial system and on to various organs and muscle beds.

In the gastrointestinal tract, the blood gathers and transports nutrients from the digestive system for distribution to other organs and muscle sites. At the muscle sites, the blood is essentially trading oxygen and nutrients in the capillary beds for carbon dioxide, water, heat, and lactic acid. Finally, the blood flows back toward the heart through the venous return system.

This description is simplified and neglects other aspects, such as waste removal and filtration through the kidneys and other organs like the skin, bone, and brain, as well as fatty tissues and vessels supplying the heart muscle, and also the lymphatic system (Kapit and Elson 2002).

The Metabolic System

Now that we have covered a basic introduction to several body systems, including the skeletal, skeletal muscular, neuromuscular, respiratory, and circulatory, we can finally address the *metabolic system*, which is critical to our study of work physiology. The metabolic system is the process by which the body consumes and produces energy for the purposes of existence and work output. What follows is a simplified version of metabolism. References for this section include Åstrand et al. (2003); Kapit and Elson (2002); Kapit, Macey, and Meisami (2000); Konz and Johnson (2004); and Kroemer, Kroemer, and Kroemer-Elbert (1997).

The Human Engine and Energy Balance

Over time, the body maintains a balance between energy input and output. Most physiology texts describe the metabolic system as the engine in the human body. The metabolic system includes elements of several systems, including the digestive, respiratory, circulatory, and muscle systems. The inputs to the metabolic system are primarily oxygen and food. The digestive system receives and processes the food. The intestines and liver pass nutrients into the bloodstream, and they are delivered to various muscle sites about the body for metabolism of energy, usually in the presence of oxygen (aerobic), but sometimes without the presence of oxygen (anaerobic). The products from this metabolism are external energy (in the form of work) and internal energy (consumed to maintain body temperature and fuel internal organs). By-products of the metabolism are carbon dioxide, water, and heat. Some of the heat and water is transported close to the skin and lost through the skin by perspiration and convective heat loss. The remainder of the carbon dioxide, water, and heat is transported back to the alveolar interface in the lungs and expired by the respiratory system (Kapit and Elson 2002; Kapit, Macey, and Meisami 2000).

The energy balance equation, which is also used in the discussion of the thermoregulatory system, is as follows (Kroemer, Kroemer, and Kroemer-Elbert 1997; Åstrand et al. 2003):

$$M \pm S \pm R \pm C \pm K - W - E = 0 \qquad (3)$$

where,

M = metabolic rate
S = heat storage rate
R = radiant heat exchange rate
C = convective heat exchange rate
K = conductive heat exchange rate
W = mechanical work rate
E = evaporative cooling exchange rate.

Units can be in watts (W) or in joules/second (J/s) (1 W = 1 J/s).

Equation 3 assumes that the person consumes exactly the same number of calories as burned through metabolism, otherwise the person would stand to lose or gain weight, or the equation would need a term for energy storage rate (lost or gained), primarily in the form of fat lost or added. In either case, the body is constantly striving for balance, or homeostasis. The metabolic rate is always positive as the body is constantly producing the energy needed for basic existence (basal metabolism: body temperature, base body functions, and blood circulation), plus what is required for current activity, and current digestive metabolism.

W is the mechanical work rate, or the external work produced. Note that W is always a loss (or zero in the resting case) from this equilibrium equation, as it is not normal to experience a gain of external work back to the system. Note also that evaporative cooling is always a loss (or zero) to the system.

S is positive if the body heat content increases, and negative for loss of body heat. Normally this number should remain close to zero, or at least be nonzero for very short periods of time, or else the body is risking hypothermia (negative) or heat stroke (positive).

Radiant heat exchange (R) would be positive due to exposure to the sun or a radiant heat source, such as a lamp or blast furnace. R can also be negative if the body is radiating heat; thus, the net R can be positive or negative. Convective heat gain (C) occurs when air warmer than the skin temperature is encountered, and loss occurs for the opposite situation. Conductive heat gain (K) occurs if the body is in physical contact with a warmer body, such as sitting on a heated surface, and heat loss occurs if seated on a cold surface.

Thermal Stress

Thermal stress is of particular interest in work physiology, as many work tasks must be carried out in extremes of heat and cold. The body's core temperature must be maintained close to 37°C (98.6°F). The key organs of the core are the brain, heart, lungs, and abdominal organs. Changes in core temperature of ±2°C affect body functions and task performance severely. Variations of ±6°C are usually lethal (Kroemer, Kroemer, and Kroemer-Elbert 1997).

Heat energy is circulated through the body by the blood. The blood flow can be controlled by vasomotor actions in the blood vessels. Vasomotor actions include *constriction* (narrowing of the blood vessels), *dilation* (expanding the blood vessels), and *shunting* (shifting flow from superficial to deep blood vessels or vice versa). In a cold environment, heat must be conserved, and in a hot environment, heat must be dissipated, while gain from the environment must be prevented. The most efficient way to dissipate heat from the body to the environment is through the skin and, to a lesser extent, through the lungs. Involuntary shivering in the muscles results in internal heat generation. Of course, clothing and shelter can provide a significant effect on thermal control. Evaporative heat loss through the skin can take place when air moves against skin covered with perspiration. To promote evaporative loss, increase the amount of exposed skin by wearing less clothing and using fans to increase air flow. Lower humidity improves evaporative performance, while high humidity can make it much less effective.

In heat stress situations, it is desirable to limit radiant heat gain from sources such as the sun, a blast furnace, or open flames. Providing shade or a barrier can limit the radiant exchange. Clothing can be an effective barrier, but will interfere with evaporative cooling. In extreme heat stress situations, ice packs can be worn near the skin, the work rate may need to be limited, or the work time may need to be restricted. In extreme cold stress situations, it is important to keep skin covered and protected. Insulated clothing layers are necessary. Extra care must be taken to protect the fingers, feet, face, and neck.

For more in-depth coverage of this topic, consult Kroemer, Kroemer, and Kroemer-Elbert (1997) and Kapit, Macey, and Meisami (2000).

Energy Requirements for Work

Some examples of the energy required for activities (in watts per kilogram, W/kg) include sitting quietly or writing (0.4), standing office work (0.7), driving a car (1.0), washing floors (1.2), sweeping floors (1.6), heavy carpentry (2.7), cleaning windows (3.0), and sawing wood by hand (6.6). For total energy cost, add basal metabolism of 1.28 W/kg for males and 1.16 W/kg for females (Konz and Johnson 2004).

Watts are the SI unit of power. *Power* is the rate at which work is done, or (equivalently) the rate at which energy is expended (Hibbeler 2001). One watt is equal to a power rate of one joule of work per second of time. In these examples, the units are given in watts per unit of body weight in kilograms.

The metabolic cost of walking can be calculated as follows (Pandolf, Haisman, and Goldman 1976):

$$\text{Walking Metabolism} = C[2.7 + 3.2(v - 0.7)1.65] \quad (4)$$

in W/kg of body weight, where C, the terrain coefficient,

= 1.0 blacktop road, treadmill
= 1.1 dirt road
= 1.2 light brush
= 1.3 hard-packed snow
= 1.5 heavy brush
= 1.8 swamp
= 2.1 sand

and v = velocity in meters/second (m/s), where $v > 0.7$ m/s.

The terrain coefficient represents the degree of difficulty presented by the walking. For example, it is much more difficult to walk in sand than to walk on a flat, hard surface. A 200-pound (90.7-kilogram) person, walking at 3 miles per hour (mph) (1.34 m/s), which is considered a normal walking speed, on a flat, hard surface (C = 1.0), would yield 4.23 W/kg of their body weight, or 384 W. A 150-pound (68.0-kilogram) person walking at 2 mph (0.89 m/s) in sand (C = 2.1) would yield 6.1 W/kg, or 415 W.

The metabolic cost of carrying a load in one's hands is 1.4 to 1.9 times the energy of carrying your own body weight (Soule and Goldman 1980).

The metabolic cost for lifting can be calculated as follows (adapted from Garg, Chaffin, and Herrin 1978):

Lifting Metabolism =
[0.024(BW) + F(Load Factor)] ÷ 0.014314   (5)

in watts, where

BW = body weight (kg)
F = frequency of lifts (lifts/min).

Load Factor = LFBW(BW) + LFL(W) + GF(W)   (6)

where LFBW, the lift factor body weight,

= 0.00044, for arm lift
= 0.00265, for stoop lift
= 0.00419, for squat lift

LFL, the lift factor load,

= 0.02271, for arm lift
= 0.01147, for stoop lift
= 0.01786 for squat lift

GF, the gender factor,

= –0.00375(G), for arm lift
= –0.00617(G), for stoop lift
= –0.00507(G), for squat lift

G, the gender,

= 0, female
= 1, male

and W is the object's weight in kilograms.

For example, a 200-pound (90.7-kilogram) male performing arm lifts of boxes weighing 25 pounds (11.3 kilograms) at a rate of 4 lifts per minute would yield 223 W. A 150-pound (68.0-kilogram) female performing squat lifts of boxes weighing 15 pounds (6.8 kilograms) at a rate of 4 lifts per minute would yield 228 W.

In everyday activities, only about 5 percent or less of the energy input is converted into work. The remainder is mostly converted into heat. In meeting work demands, the body is called to increase energy production up to 50 times that of resting state. In addition to the importance of temperature control in this equation, the body has a tremendous ability to meet the demand, largely depending on the circulatory and respiratory systems to serve the involved muscles in order to meet such a 50-fold requirement (Kroemer, Kroemer, and Kroemer-Elbert 1997).

TABLE 3

Interaction Between Aerobic and Anaerobic Processes to Meet Maximal Efforts

	Exercise Time, Maximal Effort		
Process	10 seconds	10 minutes	2 hours
Anaerobic			
kJ	100	150	65
kcal	25	35	15
Percent	85	10–15	1
Aerobic			
kJ	20	1000	10,000
kcal	5	250	2400
Percent	15	85–90	99
Total			
kJ	120	1150	10,065
kcal	30	285	2415

(Adapted from Åstrand and Rodahl 1986, p. 325)

AEROBIC AND ANAEROBIC PROCESSES

Under normal working conditions, it is extremely desirable for the body to be producing energy almost exclusively by aerobic (requiring oxygen) processes. In an aerobic task, a worker can easily take in all the oxygen required for the task with very little or no production of lactic acid in the muscle capillary beds. A good rule of thumb is that the worker should be able to talk easily while breathing during the work task.

When the body is asked to meet a demand of the greatest possible effort over short periods of time, it has the ability to meet that demand with anaerobic (without oxygen) processes, but at a very high cost. There will be a lactic acid buildup which will shut down the ability to continue to perform the effort. Buildup of lactic acid in the tissues quickly leads to muscular fatigue and a subsequent slowdown or shutdown of work. Significant buildup of lactic acid in the tissues takes considerable time to remove—one hour or more (Åstrand et al. 2003). Thus, there is a heavy penalty for performing anaerobic work, and it should be avoided for normal working conditions. The relative interaction between aerobic and anaerobic energy processes is shown in Table 3. Note that a 2-hour maximal effort uses less anaerobic energy [65 kilojoules (kJ)] than a 10-second maximal effort (100 kJ). And a 10-second maximal effort uses less aerobic energy (20 kJ) than a 2-hour maximal effort (10,000 kJ) (Kroemer, Kroemer, and Kroemer-Elbert 1997, Åstrand et al. 2003).

The practical application of this information is to normally design jobs that are almost entirely aerobic to avoid the buildup of lactic acid. Four cases, ranging from light work to extremely heavy work, are given. An example of the physiologic response to work is shown in Figure 1 and illustrates the cases of light work and moderate-intensity work. Prior to onset of work, the body is in a state of equilibrium at resting level. When the work begins, the demand to perform the work is a step function. Even if the workload is relatively light, the aerobic response cannot satisfy the step function. But the body can meet the step demand with anaerobic processes. Soon, the aerobic processes meet the demand and reach a state of aerobic equilibrium. Eventually, the utilized anaerobic energy must be returned (or paid for), much like recharging batteries for their next use:

- During *light work*, the oxygen stored in the muscle, plus the oxygen supplied from respiration and circulation, will completely cover the oxygen need (repaying the oxygen deficit) (see Figure 1).
- During *work of moderate intensity*, anaerobic processes contribute to the energy output at the beginning of the task until aerobic processes can take over and completely cover the energy demand. The lactic acid produced diffuses into

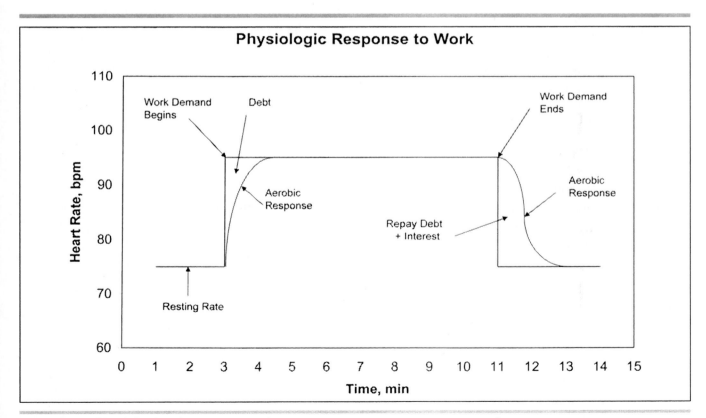

FIGURE 1. Physiologic response to work
(Adapted from Kroemer, Kroemer, and Kroemer-Elbert 1997, p. 217)

the blood, but the concentration returns to resting level and the work task can continue for hours (see Figure 1).
- During *heavier work*, the lactic acid production and concentration in the blood rises and remains high. The length of time the work rate can continue will be due largely to the motivation level of the worker.
- During *extremely heavy work*, the lactic acid concentration grows significantly, and the oxygen deficit cannot be recovered. The work task cannot continue for more than a few minutes, and it may take up to 60 minutes for the lactic acid concentration to recover to resting levels (Åstrand et al. 2003). When work tasks of this magnitude are required, it is necessary to implement work teams and generous work-rest ratios.

Part of this discussion centers on *demand versus capability*. As long as the aerobic capability of the worker exceeds the aerobic demand of the work task, suitable tasks can be designed for continuous, eight-hour shifts. A practical approach is to apply the rule of thumb that suggests workers should be able to talk easily while breathing. Otherwise, the difficult task of estimating or measuring task requirements and estimating the aerobic capacity of the range of workers is required. This issue will be discussed in a later section.

The cardiovascular system responds to work demand or exercise with changes in heart rate, heart stroke volume, artery-vein differential, and blood distribution, and by going into debt (as was just discussed). In response to increased workload or work demands, the body begins to accommodate and adjust by increasing the heart rate and stroke volume, thus delivering more oxygen to the muscle sites for metabolism. With the artery-vein differential response, the body takes more oxygen out of each unit of blood. Under resting conditions, the body only uses a portion of the available oxygen, therefore there is potential to use more when necessary. With the blood distribution response, the body has the ability to divert blood from areas of less critical immediate need to areas of greater need (see Table 4). Although the digestive system, kidneys, bone, and brain receive a smaller percentage of the blood volume during heavy work, the net volume of blood to these systems remains very similar to that supplied at rest because the cardiac output is up to five times greater during heavy work (Kroemer, Kroemer, and Kroemer-Elbert 1997; Åstrand et al. 2003; Kapit, Macey, and Meisami 2000).

EVALUATION OF CARDIOVASCULAR CAPACITY
Oxygen Uptake or Consumption

It would seem logical to evaluate the cardiovascular capacity of employees to determine safe and effective job placement. (Note that screening must be done carefully to maintain compliance with Federal Equal Opportunity Employment Laws and the Americans with Disabilities Act.) An individual's cardiovascular capacity can be determined by measuring maximum oxygen uptake, $\dot{V}O_2$max—or the rate at which one can process oxygen at maximal exertion (Åstrand et al. 2003). But this is not a simple measurement to collect. On an absolute basis, the units for maximum oxygen uptake, $\dot{V}O_2$max, are in liters per minute (L/min); on a relative basis (with respect to body weight), the units for maximum oxygen uptake are in milliliters per kilograms per minute [mL/(kg/min)]. There are instruments that measure the concentration of oxygen exhaled as well as the volume of breathing air. This is normally done in an exercise physiology laboratory,

TABLE 4

Distribution of Blood at Rest and During Heavy Work

Organ/System	At Rest (%)	During Heavy Work (%)
Lungs/heart	100	100
Digestive system	20–25	3–5
Cardiac arteries	4–5	4–5
Kidneys	20	2–4
Bone	3–5	0.5–1
Brain	15	3–4
Fatty tissue	5–10	1
Skin	5	80–85
Muscle	15–20	

Note: Numbers are estimated percentages. Due to increased heart rate and stroke volume, the cardiac output can be five times greater when changing from rest to heavy work.

(Adapted from Åstrand and Rodahl 1986, p. 152)

but there are portable instruments available for field studies (for example, Oxylog: PK Morgan, Kent, UK). Even the portable instruments need calibration and present a bit of an obstacle in the workplace. Fortunately, there is a relatively linear relationship between oxygen uptake (consumption) and heart rate during work, so the exercise rate can be used to estimate $\dot{V}O_2max$ (Kroemer, Kroemer, and Kroemer-Elbert 1997, Åstrand et al. 2003) (see Table 5).

The formulas for calculating maximum heart rate were previously given. The formula for calculating the percent of maximum heart rate (Equation 7 shows an example at 50 percent), requires an estimate of resting heart rate, which can be easily obtained (Burke 1998):

50% of Max HR =
[(Max HR − Resting HR)(0.5)] + Resting HR   (7)

For example, a 35-year-old worker has a resting heart rate of 65 bpm. Using Equation 1, we can estimate the maximum heart rate (Max HR) at 185bpm. Therefore, using Equation 7, 25 percent of Max HR is approximately 95 bpm, 50 percent is approximately 125 bpm, and 75 percent is approximately 155 bpm.

There are maximal and submaximal stress tests that can be used to estimate $\dot{V}O_2max$. They each have their advantages and disadvantages. The treadmill tests have an advantage in that it is more of a whole-body test, and of course the individual must carry their entire body weight throughout the test. The speed and incline are dictated by the protocol.

The cycle ergometer tests have some advantages in that the instrument is more portable and does not require power. The ergometer test is more independent of body weight if that is desirable. The cadence and resistance is governed by the protocol. Instrumentation, such as heart rate monitors and face masks, are more stable with the ergometer.

The step tests are the most portable and may be best suited for field studies. The step height is dictated by the protocol. The step rate is governed by a metronome in an up-up-down-down sequence. The individual must carry their entire body weight. Note that with a fast cadence and fatigue setting in, the step test presents a trip hazard (Åstrand et al. 2003).

TABLE 5

Relationship Between Relative Heart Rate and Percent $\dot{V}O_2max$

Percent Max HR	Percent $\dot{V}O_2max$
35	30
60	50
80	75
90	84

Note: Regression equation for this data: Percent $\dot{V}O_2max$ = 1.07(Percent Max HR) − 10.2   (8)

(Source: E. R. Burke (ed.), Precision Heart Rate Training, 1998, p. 19)

The maximal tests are the most accurate, but require a trained staff (including emergency-trained medical personnel), more sophisticated laboratory equipment, and are more time consuming. The test is performed on a treadmill or cycle ergometer, and the individual follows a test protocol that takes them to the point of complete exhaustion or collapse. For example, the maximal treadmill test calls for an increase in speed and slope every three minutes until even a world-class athlete reaches a collapse point. The maximum heart rate and $\dot{V}O_2max$ are measured directly just prior to collapse.

A set of submaximal stress tests are available for the treadmill, cycle ergometer, or raised steps. These tests are designed as submaximal and, by following the prescribed protocols, are inherently safer, especially for older or less-fit individuals. Heart rate data are sufficient, thus oxygen concentration and lung volume instrumentation is not needed. Laypersons can administer the test without medical personnel present. The test is essentially searching for the equilibrium heart rate required by the individual to perform at a fixed submaximal level. That equilibrium rate alone can be used to monitor the fitness of an individual performing the same test longitudinally. The equilibrium rate can also be used to extrapolate to the $\dot{V}O_2max$. It is not as accurate as the maximal test.

Two final field tests are presented that are quite simple in terms of equipment. All that is required is a timing device and a track or measured distance. Because the pace of these tests is not controlled, emergency-trained medical personnel should be present and a physician's consent for older or at-risk individuals should be obtained.

- Measure the distance a person can walk or run in 12 minutes (Konz and Johnson 2004):

$$\dot{V}O_2max = -10.3 + 35.3 (DIST) \quad (8)$$

- In mL/(kg/min)
- DIST, miles covered in 720 s (12 min)

For example, a person walks 1.0 mile in 12 minutes. Thus, their estimated $\dot{V}O_2max$, using Equation 8, is approximately 25 mL/(kg/min). If a person covers 1.5 miles in 12 minutes, their $\dot{V}O_2max$ is approximately 43 mL/(kg/min).

- Record the time required for a person to walk or run 2 kilometers (km) (Bunc 1994):
  - In mL/(kg/min)
  - $\dot{V}O_2max = 85.7 - 251.3(T)$ [for males] (9)
  - $\dot{V}O_2max = 61.9 - 124.2(T)$ [for females] (10)
  - T = time to run 2 km (hr)

For example, a man and a woman each cover the 2-kilometer distance in 10 minutes (0.17 hour). Their approximate $\dot{V}O_2max$, using Equations 9 and 10, respectively, is 43 and 41 mL/(kg/min).

In order to help evaluate the maximum oxygen uptake, some data for U.S. males are presented in Table 6. Females typically have a $\dot{V}O_2max$ 15–30 percent below males (Konz and Johnson 2004). If you compute absolute $\dot{V}O_2max$, in L/min and want relative $\dot{V}O_2max$ in mL/(kg/min), divide the absolute by the individual's body weight in kilograms.

The numbers in Table 6 represent a cross-section of U.S. males and are estimated for U.S. females at 75 percent of the male data.

Conversion of Oxygen Consumption to Energy and Energy Conversion

For *normal* adults on a *normal* diet doing *normal* work, we can calculate the energy conversion occurring in the body from the volume of oxygen consumed. Considering a *normal* nutritional diet of carbohydrates, fats, and proteins, the overall *average* caloric value of oxygen is 5 kilocalories per liter of oxygen (kcal/L O_2) or 21 kilojoules per liter of oxygen (kJ/L O_2). The energy conversion is fairly complex, but here are the necessary conversion factors:

$$1 \text{ L } O_2 = 5 \text{ kcal} = 21 \text{ kJ} \quad (11)$$

$$1 \text{ kcal} = 1 \text{ cal} = 1000 \text{ cal} \quad (12)$$

$$1 \text{ W} = 1 \text{ J/s} \quad (13)$$

$$1 \text{ J} = 1 \text{ Newton-meter (Nm)} = 0.239 \text{ cal} = 0.000239 \text{ kcal} = 10^7 \text{ ergs} = 0.938 \times 10^{-3} \text{ Btu} = 0.7376 \text{ ft-lb} \quad (14)$$

$$1 \text{ W} = 0.85885 \text{ kcal/hr} = 0.014314 \text{ kcal/min} \quad (15)$$

Example

In an earlier discussion we learned that heavy carpentry required 2.7 W/kg + 1.28 W/kg for basal metabolism for males. Let us convert that to mL/(kg/min) of relative oxygen consumption: 3.98 W/kg (0.85885 kcal/hr per W) = 3.418 kcal/(kg/hr) = 0.057 kcal/(kg/min). Then multiply that by 1 L O_2/5 kcal = 0.01139 L O_2/(kg/min). = 11.39 mL O_2/(kg/min).

TABLE 6

Cardiovascular Fitness	Maximum Oxygen Consumption $\dot{V}O_2max$, in mL/(kg/min)							
	Male Age (Years)				Female Age (Years)			
	< 30	30–39	40–49	≥ 50	< 30	30–39	40–49	≥ 50
Very Poor	< 25	< 25	< 25	--	< 19	< 19	< 19	--
Poor	25–34	25–30	25–26	< 25	19–26	19–23	19–20	< 19
Fair	34–43	30–39	26–35	25–34	26–32	23–29	20–26	19–26
Good	43–52	39–48	35–46	34–43	32–39	29–36	26–35	26–32
Excellent	> 52	> 48	> 46	> 43	> 39	> 36	> 35	> 32

(Adapted from Cooper 1970, p. 28)

Let us assume we expect a person to work at that pace for an 8-hour shift. We are soon to learn that in the design of work, it is recommended an 8-hour shift not exceed 33 percent of $\dot{V}O_2$max. If we set 11.39 mL/(kg/min) as 33 percent of maximum, then the maximum would be 34.5 mL/(kg/min). Referring to Table 6, it would appear that males less than 30 years old would need to have *fair* or better cardiovascular fitness and those older than 30 would need to have, at least, *good* cardiovascular fitness to perform heavy carpentry work over an 8-hour shift and be able to avoid anaerobic metabolism and buildup of lactic acid. For females, accounting for a slightly lower basal metabolism, the maximum would be 33.5 mL/(kg/min). It would appear that females less than 30 years old would need to have *good* cardiovascular fitness or better and those older than 30 would need to have, at least, *excellent* cardiovascular fitness to avoid anaerobic metabolism for 8-hour shifts in heavy carpentry.

Classification of Work

Energy requirements for a job allow for judgment to be used to classify whether a job is easy or hard. Although these judgments rely on various factors, one such classification is given by Kroemer, Kroemer, and Kroemer-Elbert (1997) in Table 7. These measurements represent work performed over a whole work shift according to energy expenditure and heart rate.

TABLE 7

Classification of Light to Heavy Work

Classification	Total Energy Expenditure				HR (bpm)
	in kJ/min	in kcal/min	W	W/kg[1]	
Light work	10	2.5	174	2.6	≤ 90
Medium work	20	5.0	349	5.1	100
Heavy work	30	7.5	523	7.7	120
Very heavy work	40	10.0	697	10.3	140
Extremely heavy work	50	12.5	872	12.8	≥ 160

[1] For a 150-lb (68-kg) person.

(Adapted from K. H. E. Kroemer, H. J. Kroemer, and K. E. Kroemer-Elbert 1997, p. 222)

FATIGUE

A discussion of *fatigue* with regard to work physiology is essential. Some discussion of fatigue occurred earlier in this chapter. The following sections discuss localized muscle fatigue, general physiologic fatigue, and mental fatigue. Everyone experiences fatigue at some point during the day. It can be mental, physical, or both. It is usually accompanied by a loss of efficiency.

Localized Muscle Fatigue

Localized muscle fatigue was discussed in some detail earlier in the coverage of the skeletal muscular system. Prolonged isometric (static) muscle exertions should be avoided. In these cases, the muscle motor units are over-used and the circulatory system is unable to provide oxygen and nutrients to the muscle cells, and is unable to remove carbon dioxide and lactic acid as well (Chaffin, Andersson, and Martin 2006). While prolonged static work should be avoided, it cannot be avoided altogether. All work tasks contain at least some elements of static work.

For example, while standing, several muscle groups in the legs, hips, back, and neck are tensed to hold that position. Standing tasks that include walking are much more comfortable because walking relieves the tension and ensures blood flow through the muscle tissues. When sitting down, the muscle tension in the legs, hips, and back are greatly reduced. However, prolonged sitting with little motion does not promote a healthy flow of blood through the muscle tissues. When lying down, almost all muscle tension is avoided. A recumbent position is the most restful and, as we know, is the best position for sleeping.

In summary, a standing/walking task is probably healthiest and least likely to produce localized muscle fatigue. People who walk for a living need to be given a chance to sit occasionally and be encouraged to wear comfortable, supportive footwear. People who stand for a living must be encouraged to move very frequently, sit occasionally, wear comfortable, supportive footwear, and, where practical, use anti-fatigue mats. People who sit for a living should use supportive, adjustable chairs, and should be encouraged to walk about frequently.

Static effort should be addressed if the following circumstances exist (Kroemer and Grandjean 1997):

- a high level of effort is maintained for 10 seconds or more
- a moderate effort persists for 1 minute or more
- a slight effort (about one-third of maximum force) lasts for 5 minutes or more

General Physiologic and Mental Fatigue

In addition to localized muscle fatigue, there are other types of general physiologic fatigue and mental fatigue that can reduce or impair performance and lead to a general sensation of weariness. A feeling of weariness is not always unpleasant, especially if it is time to sleep or relax. It can be a protective mechanism to discourage one from overstraining and encourage rest and recuperation. In fact, rest and recuperation is the great balancing factor that acts to offset stress and fatigue. Recuperation takes place mainly during night-time sleep, but free periods during the day and all kinds of pauses during work also make their contributions. Some other types of somewhat distinguishable physiologic fatigue are (Kroemer and Grandjean 1997):

- eye fatigue caused by straining the visual system
- general body fatigue—physical overloading of the entire organism
- mental fatigue induced by mental or intellectual work
- nervous fatigue caused by overstressing one part of the psychomotor system, as in skilled, often repetitive, work
- chronic fatigue, an accumulation of long-term effects
- circadian fatigue, part of the day-night rhythm and initiating a period of sleep

Some control measures for general physiologic fatigue that can be employed are:

- First, recognize the symptoms of fatigue. This is a responsibility for managers as well as workers.
- Realize that extended periods of overtime may help boost production, but possibly at a cost to the well-being of the workforce and to the quality of production.
- Work with ergonomists, safety engineers, industrial engineers, and industrial hygienists to ensure a sound design of work practices, work stations, tools, and the overall environment. Provide variety in the design of work. Use principles of job enlargement and job enrichment.
- Encourage worker participation. Use employee suggestion systems. Collect work feedback. Solicit workers for solutions to problems. Workers who participate in solving problems are more likely to accept changes in the workplace.
- Pay attention to lighting design. Fluorescents are generally over-used and can be harsh. Make use of indirect lighting where possible. Be vigilant for unnecessary glare on computer monitors. Tilt monitors down a few degrees to eliminate most glare. Align monitor surfaces perpendicular to external windows—never parallel.
- Provide comfortable break rooms that are separated from the work area. Maintain a comfortable temperature and humidity. Use indirect lighting. Provide an ample supply of fresh, cool water and clean, easily accessible restrooms. Provide healthy food or snack alternatives, including protein.
- Train for and promote the concept of microbreaks. These are short pauses of less than a minute that are integrated into the workday, but do not significantly affect productivity. During these microbreaks, workers perform strategic stretching and strengthening exercises that promote blood flow and disrupt localized muscle fatigue.
- Provide regular employee training programs that discuss symptoms and causes of stress and fatigue. Educate employees on pertinent risk factors, with the realization that some of these risk factors come not only from the

workplace, but from hobbies, home life, and personal interrelationships.
- Provide access to health and fitness activities, or at least encourage participation.
- Provide employee assistance programs for those coping with financial, emotional, and substance abuse difficulties.

WORK DESIGN

What is the best way to design work? How much should workers be physically challenged? Should work be designed for gender or age differences? How should work schedules be designed? What effects do special work environments have (thermal stress, noise, and so on)? How can companies continue to meet the challenges of increasing productivity and improving quality?

There are ample opportunities in work design to address these issues. One of the major reasons for optimism is that most of these issues are misunderstood or ignored. Another reason for optimism is that addressing these issues will often lead to significant cost savings.

Design for Work Capacity

What portion of a person's overall physical work capacity is reasonable from a design standpoint? It was pointed out earlier that it is desirable to avoid anaerobic metabolism as much as possible. Here are some suggested guidelines. Note that the authors do not all agree, but together they do give a general sense of a reasonable work design level. And just because workers can be pushed to these levels, does not mean they should be:

- 50 percent of maximum capacity for trained workers, 33 percent for untrained workers; reduce the level by 30 percent if the task is primarily upper-body work (Jorgensen 1985)
- set lifting limits at 21–23 percent of uphill treadmill aerobic capacity or 28–29 percent of bicycle aerobic capacity, reduce to 23–24 percent of bicycle aerobic capacity for 12-hour shifts (Mital, Nicholson, and Ayoub 1993)

- 33 percent of maximum capacity for an 8-hour shift, 35.5 percent for a 10-hour shift, 28 percent for a 12-hour shift (Eastman Kodak 1986)
- 43–50 percent of maximum capacity for package handling for a 2-hour shift (Mital, Hamid, and Brown 1994)
- assuming you want to exclude only a small percentage of the population, set your limits at about 350 W, 5 kcal/min, and 100–120 bpm (Konz and Johnson 2004)
- do not exceed 110 bpm generally for a working day; do not exceed 130 bpm for intensive work periods (Wisner 1989)
- for extended periods of work, do not exceed a maximum of 35 bpm over resting level (Kroemer and Grandjean 1997)

Reduce cardiovascular stress first with engineering solutions, then administrative solutions. High metabolic rate jobs are prime candidates for mechanization. For material handling, consider use of conveyors, hoists, and forklifts. Workers should slide or lower objects rather than lift them. Use wheeled carts rather than carrying loads. Use powered handtools. Balancers, manipulators, and jigs can reduce static loads. Administrative solutions include job rotation and part-time work. When setting work standards, apply reasonable fatigue allowances (Konz and Johnson 2004).

Design for Gender and Age Differences

There are physiological differences between males and females. Females have a $\dot{V}O_2$max 15–30 percent lower than males on average, largely because of a higher percent of body fat and a lower hemoglobin level. Females also have lower blood volumes and lung volumes on average. Rather than designing jobs differently for each gender, design jobs at a reasonable percentage of maximum for females, and males will be accommodated also (Konz and Johnson 2004).

The physiological peak performance age is approximately between ages 25 and 30. After age 30, there is a steady decline in physiological performance. A decline in $\dot{V}O_2$max is estimated at 1–2 percent per year after age 25, but there are large individual variations (Illmarinen 1992). Jackson, Beard, Weir, and Stuteville (1992) estimate the aging effect at 0.27 mL/(kg/min) per year, but emphasize that most of the decline is due to physical activity level and body fat and not actually aging. This implies that companies are wise to promote fitness and wellness programs. From a work design standpoint, light to moderate physical work is not sensitive to aging up to about age 65. But hard, exhausting work is strongly age-dependent, with a maximum capacity between ages 20 and 25 (Konz and Johnson 2004).

Work Schedules and Circadian Rhythm

Shift work is popular because companies want to maximize the use of machines and production facilities, as well as provide customer service 24 hours a day, 7 days a week. Many shift alternatives have developed in addition to the traditional work week of 8 hours a day, 5 days a week. Alternative compressed work weeks have become commonplace:

- 4 days of 10 hours/day
- 4 days of 9 hours/day, plus 4 hours on Friday
- 4 days of 9 hours/day in week 1, then 5 days of 9 hours/day in week 2
- 3 days of 12 hours/day in week 1, then 4 days of 12 hours/day in week 2

The advantages to these schedules are longer weekends and fewer commutes. A disadvantage is shorter overnight recovery times. As discussed earlier, the body needs adequate muscle recovery time to rest and reduce overall fatigue.

Overtime is frequently used to extend the work week to meet increased production demand. Many hourly employees are motivated to work the overtime to increase their income. But overtime cuts into recovery time, with possible poor consequences for muscle recovery and fatigue. Short periods of overtime are acceptable, but prolonged overtime will begin to lose its advantages to the potential increases in fatigue, injuries, and workers' compensation claims. Part-time workers, temporary workers, and job-sharing can provide some relief for this issue.

Circadian rhythm is the body's internal cycle that lasts about 24 hours (between 22 to 25 hours) and has natural variations for numerous body functions, including body temperature, heart rate, blood pressure, respiratory volume, adrenalin production, mental abilities, release of hormones, and melatonin production. Cortisol (the "wake-up" hormone) peaks around 9 AM, and melatonin (the "go-to-sleep" hormone) peaks around 2 AM.

The most important function geared to circadian rhythm is sleep and, for most people, a normal pattern is to sleep at some stretch during the night and generally be awake and active during the day. Challenges arise when workers are asked to work at times that disrupt the circadian rhythm, such as work shifts during the evening hours or during the night-time hours. In general, the human organism is performance-oriented during the daytime and ready for rest at night (Kroemer and Grandjean 1997). Here are some of the challenges:

- Workers on permanent evening or night shifts struggle in attempting to adjust to a different schedule during the weekend so that they may function and socialize with family and friends.
- Evening or night-shift workers may have difficulties finding regular eating times and finding quiet, dark places to sleep during the daytime.
- Some companies rotate shift workers so that (for example) every three weeks the evening shift rotates to days, the night shift rotates to evenings, and the day shift rotates to nights.
- Some workers (for example, military personnel, doctors, and security guards) are on-call for 24-hour shifts.
- Some workers fly across several time zones and land in a greatly altered daily routine.

It is easier to describe the challenges surrounding shift work and circadian rhythms than it is to solve the challenges and problems. Naps can be helpful, but the length and timing of the naps is controversial. One long night's sleep usually restores performance to a normal level, even after extensive sleep deprivation. During prolonged work, periods of vigorous exercise and fresh air may help maintain alertness. If shifts are necessary, either work only one evening or night shift per cycle, then return to day work and keep weekends free; or stay permanently on the same shift, whatever that is (Kroemer, Kroemer, and Kroemer-Elbert 1997).

SUMMARY

This chapter provides the reader with an introduction to work physiology and anthropometry. Since this work is not intended to be a comprehensive treatment of these topics, readers should consult the references and other comprehensive works in these fields. Safety professionals are encouraged to learn more about how the human body is constructed and how it functions in the work environment, in order to be better able to evaluate existing work environments and to properly design new work environments. The safety professional's dual goals are to provide a safe and productive workplace.

REFERENCES

Abraham, S., C. L. Johnson, and M. F. Najjar. 1979. "Weight and Height of Adults 18–74 Years of Age, United States, 1971–1974." *Vital and Health Statistics*. Series 11, No. 211, PHS 79-1659. MD: U.S. Department of Health, Education and Welfare.

Åstrand, P.-O., and Kaare Rodahl. 1986. *Textbook of Work Physiology: Physiological Bases of Exercise*. 3d ed. New York: McGraw-Hill.

Åstrand, P.-O., Kaare Rodahl, Hans A. Dahl, and Sigmund Stromme. 2003. *Textbook of Work Physiology*. 4th ed. Champaign, IL: Human Kinetics Publishers, Inc.

Bunc, V. "A Simple Method for Estimating Aerobic Fitness." 1994. *Ergonomics* 37(1):159–65.

Burke, E. R., ed. 1998. *Precision Heart Rate Training*. Champaign, IL: Human Kinetics Publishers, Inc.

Chaffin, D. B., G. B. J. Andersson, and B. J. Martin. 2006. *Occupational Biomechanics*. 4th ed. New York: John Wiley & Sons, Inc.

Collins English Dictionary, Complete and Unabridged. 2010. 10th ed. (retrieved November 30, 2010) www.dictionary.reference.com

Cooper, K. 1970. *The New Aerobics*. New York: Bantam Books.

Dalley, A. F., and A. M. R. Agur. 2004. *Grant's Atlas of Anatomy*. 11th ed. Philadelphia, PA: Lippincott Williams & Wilkins.

Eastman Kodak. 1986. *Ergonomic Design for People at Work: Volume 2*. New York: Van Nostrand-Reinhold.

Garg, A., D. B. Chaffin, and G. Herrin. 1978. "Prediction of Metabolic Rates for Manual Material Handling Jobs." *American Industrial Hygiene Association Journal* 39:661–74.

Hibbeler, R. C. 2001. *Engineering Mechanics: Statics & Dynamics*. 9th ed. Upper Saddle River, NJ: Prentice Hall.

Illmarinen, J. 1992. "Design for the Aged With Regard to Decline in Their Maximal Aerobic Capacity: Part II The Scientific Basis for the Guide." *International Journal of Industrial Ergonomics* 10:65–77.

Jackson, A., A. Beard, L. Wier, and A. Stuteville. 1992. "Multivariate Model for Defining Changes in Maximal Physical Working Capacity of Men, Ages 25 to 70 Years." In proceedings of the Human Factors Society, pp. 171–74.

Jenkins, D. B. 2008. *Hollinshead's Functional Anatomy of the Limbs and Back*. 9th ed. Philadelphia, PA: W. B. Saunders Company.

Jorgensen, K. 1985. "Permissible Loads Based on Energy Expenditure Measurements." *Ergonomics* 28(1):365–69.

Kapit, W., and L. M. Elson. 2002. *The Anatomy Coloring Book*. 3d ed. San Francisco, CA: Benjamin Cummings.

Kapit, W., R. I. Macey, and E. Meisami. 2000. *The Physiology Coloring Book*. 2d ed. San Francisco, CA: Addison Wesley Longman.

Konz, S., and S. Johnson. 2004. *Work Design: Occupational Ergonomics*. 6th ed. Scottsdale, AZ: Holcomb Hathaway.

Kroemer, K. H. E., and E. Grandjean. 1997. *Fitting the Task to the Human: A Textbook of Occupational Ergonomics*. 5th ed. Bristol, PA: Taylor & Francis.

Kroemer, K. H. E., H. J. Kroemer, and K. E. Kroemer-Elbert. 1997. *Engineering Physiology: Bases of Human Factors/Ergonomics*. 3d ed. New York: Van Nostrand Reinhold.

———. 2001. *Ergonomics: How to Design for Ease and Efficiency*. 2d ed. Upper Saddle River, NJ: Prentice Hall.

Marras, W., and J. Kim. 1993. "Anthropometry of Industrial Populations." *Ergonomics* 36(4):371–78.

Mital, A., F. Hamid, and M. Brown. 1994. "Physical Fatigue in High and Very High Frequency Manual Material Handling: Perceived Exertion and Physiological Factors." *Human Factors* 36(2):219–31.

Mital, A., A. Nicholson, and M. M. Ayoub. 1993. *A Guide to Manual Material Handling*. London: Taylor and Francis.

Nims, D. K. 1999. *Basics of Industrial Hygiene*. New York: John Wiley & Sons, Inc.

Pandolf, K., M. Haisman, and R. Goldman. 1976. "Metabolic Energy Expenditure and Terrain Coefficients for Walking on Snow." *Ergonomics* 19:683–90.

Pheasant, S., and C. M. Haslegrave. 2006. *Bodyspace: Anthropometry, Ergonomics, and the Design of Work*. 3d ed. Boca Raton, FL: Taylor & Francis.

Roebuck, Jr., J. A., 1995. *Anthropometric Methods: Designing to Fit the Human Body*. Santa Monica, CA: Human Factors and Ergonomics Society.

Soule, R., and R. Goldman. 1980. "Energy Cost of Loads Carried on the Head, Hands, or Feet." *Journal of Applied Physiology* 27:687–90.

Tayyari, F., and J. L. Smith. 1997. *Occupational Ergonomics: Principles and Applications*. New York: Chapman & Hall.

Warfel, J. H. 1985. *The Extremities: Muscles and Motor Points*. 5th ed. Philadelphia, PA: Lea & Febiger.

Wisner, A. 1989. "Variety of Physical Characteristics in Industrially Developing Countries—Ergonomic Consequences." *International Journal of Industrial Ergonomics* 4:117–38.

Principles of Human Factors

4

Steven F. Wiker

LEARNING OBJECTIVES

- Gain a workable knowledge of the field of Human Factors Engineering (HFE).

- Understand the bases for design-induced accidents that are often mistakenly assigned to human error.

- Understand the scope and usefulness of HFE methods in preventing or reducing safety problems through improved design.

- Be able to describe a general process for avoiding perceptual, cognitive, and motor-related design flaws.

- Learn basic computations and models for use in HFE design and design review.

- Be able to demonstrate the need for HFE guidance in all stages of design.

- Learn how to access sources of information for HFE design guidance and further learning.

HUMAN FACTORS ENGINEERS (HFEs) practice in an interdisciplinary field of natural, physical, social sciences, and engineering that studies the performance capabilities and limitations of humans and applies that knowledge to the design of environments, machines, equipment, tools, and tasks to enhance human performance, safety, and health. Outside the United States, no formal differentiation is made between HFE and ergonomics, making this presentation a subset of ergonomics. There are a number of variants of HFE that include engineering psychology, human engineering, and so forth. Human factors practitioners are typically trained in experimental psychology; however, there are a number of academic fields that produce HFEs today.

Human factors engineers typically work on a variety of design issues and problems that focus on human–machine–task–environment system safety. One of the exciting aspects of human factors engineering is that development and application of such principles is continually challenging. The interplay of management demands, operating environments, task design, equipment design, user characteristics and objectives influences the scope and approach for development and application of human factors engineering design principles. That said, one should systematically apply general human factors engineering principles and strategies to improve design performance and safety or to recognize when greater levels of expertise are required to handle the problem at hand.

Nearly all definitions of the safety process address the need to recognize the existence of hazards, or states of danger, in order to best use available information and resources either to eliminate or to mitigate such hazards. The overarching goal of elimination or mitigation is to reach a level of safety that avoids occurrences of injury or damage, or loss of life or property, diminishing levels of occurrence to operational or societal levels of acceptability. This chapter addresses the importance of considering HFE design principles and processes that must be addressed if interactions among humans, tasks, jobs, equipment, tools, products, and environments are to be both safe and functional.

ORIGINS AND GOALS OF HFE

There is a mistaken notion that HFE is a comparatively new field of endeavor in need of time for experimentation, maturation, and acceptance of design principles and guides. It is true that the professional or organizational origins of human factors only date to the end of World War II; however the methods used by HFE have their roots in fundamental natural and physical sciences such as psychology, mathematics, engineering, and biological sciences, predating the period of the Industrial Revolution. As has the field of safety, HFE has continued to develop theoretically, empirically, and heuristically throughout the development of humankind. Significant strides continue to be made in HFE with the advent of new technologies.

Historically, careful and systematic etiological study of accidents revealed design errors in one or more loci of the task-human-environment-machine (THEM) system. Errors directly or indirectly responsible for accidents were due to poor human factors engineering. Many of the problems found were attributable to (a) a lack of understanding on the part of the designer or design team of basic human operator constraints; (b) the failure to consider how operator performance demands changed when system failures occurred; and (c) changes in tasks, operating environments, or personnel as designs or demands evolved over time. Many good initial designs were compromised by subsequent *add-on* components, tasks, or expansions of the scope of operation without adequate HFE evaluation or testing.

For these reasons, HFE has continued to broaden its area of focus and activity to include transportation, architecture, environment design, consumer products, electronics/computers, energy systems, medical devices, manufacturing, office automation, organizational design and management, aging, farming, health, sports and recreation, oil-field operations, mining, forensics, education, speech synthesis, and many, many other arenas. Today, nearly all large corporate and military–industrial entities pair design engineers with human factors engineers to develop more usable and safe designs.

Why Do HFE Design Problems Affect Human Safety?

The Bureau of Labor Statistics (BLS) of the United States Department of Labor catalogues death rates for occupations and publishes rankings of jobs that present the greatest risk of death in the United States. In 2006, 5703 people died on the job in the United States (BLS 2007). Figure 1 shows the data graphically.

Closely associated with death rates are rates of equipment damage, facility damage, personnel selection and training costs, and other insidious costs. Because many of these accidents are attributed to human error, it is natural to ask why such problems exist. In-depth study of such accidents usually demonstrates a mismatch of the performance demands of equipment, environments, products, jobs, tasks, or system process designs with human capabilities and limitations. Most problems are originally *designed in* or created by ad hoc changes made to designs without careful design review and approval.

Five engineering design fallacies have been associated with use of anthropometric data (Pheasant 1988). Designers often suffer from one or more of the following fallacies:

- Because this design is satisfactory for me, it will be satisfactory for everybody else.
- Because this design is satisfactory for the average person, it will be satisfactory for everybody else.
- Human beings vary so greatly from each other that their differences cannot possibly be accommodated in any one design—but because people are so wonderfully adaptable, it doesn't matter anyway.

- Because ergonomics is expensive, and because products are purchased for their appearance and styling, ergonomic considerations may be ignored.
- Ergonomics is an excellent idea. It is good to design with ergonomics in mind, but do it intuitively and rely on common sense rather than tables of data.

Many designers believe that human factors engineering is simply the process of using one's *common sense*. One definition of common sense is "sound, ordinary sense (or good judgment) shared by a group at large." Unfortunately, history is replete with examples of highly educated and otherwise very successful individuals or groups who in retrospect appear to have failed to exercise or understand common sense. Moreover, a design rationale that consistently satisfies the *majority* may be inadequate or unsafe for the minority.

A number of studies have shown that multiple population stereotypes exist for any given design and

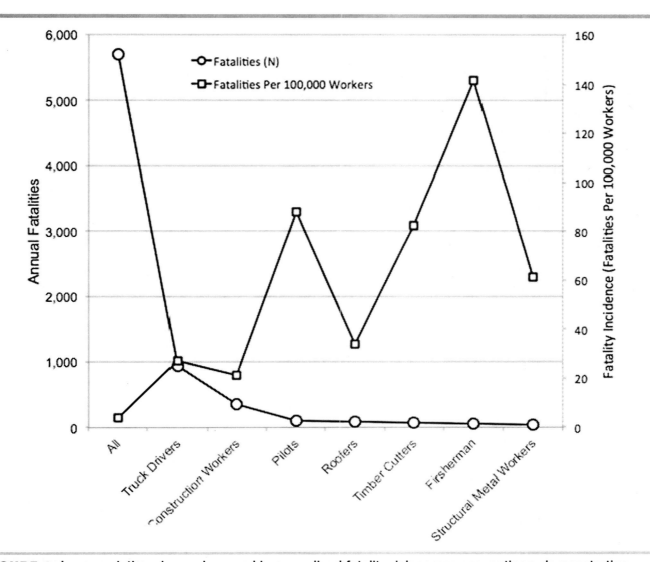

FIGURE 1. Large variations in numbers and in normalized fatality risk across occupations, demonstrating both efficacy and need for additional efforts in safety (*Source:* BLS 2007)[1]

[1] The BLS Web site should be referenced to examine the old and new reporting methodologies and fatality and injury incident information for a vast collection of occupations and exposures in the past and for future periods.

that the minority perspective can be held by large segments of the human population. Thus, one cannot presume that one's personal view of common sense matches that of the majority of individuals who are intended users of the design. In fact, it can be argued that designers and engineers who create new designs or modify existing ones are bright individuals who often "think outside of the box" to achieve breakthroughs by means of approaches, analyses, or resources used in unusual manners. These professionals are typically individuals having unique education, training, experience, and skills who by their nature are unlikely to work in the realm of common sense.

If product designs rely on individual designers' intuition or *common sense*, accidents and related problems will likely arise from the mismatch of designers' and users' mental models, expectations, and performance capabilities. To overcome this problem, it must be realized that individuals vary in their knowledge, belief systems, training, experience, skills, and their physical and mental capacities to recognize hazards, make accurate or relevant judgments, and take an appropriate corrective protective action without error.

Many people are frustrated by designs of television remote controls, by cellular phones that rely on the memorized use of multimodal button sequences, by electronic calculators, and by equipment and processes believed by their designers to follow common sense. A classic example of this problem is demonstrated by a calculator manufacturer that believed key entry should reflect the operational sequences that the computer actually used—but the key-entry sequence was not consistent with sequences followed by algebraic operations. Many customers found the number and operational keying sequence confusing, even though the designers felt the customer population would eventually recognize the benefits of using nonalgebraic key sequences.

Other designers, recognizing that a large segment of the customer base preferred to enter information into calculators using standard algebraic sequence, captured a large segment of the market by designing calculators that met those users' expectations. Today, many calculators allow the user to select either mode of data or operation.

Engineers or designers who expect users to understand their designs or share their mental model of the application and scope of use of their designs are increasingly surprised by the lack of user agreement. Matters do not seem likely to improve in light of the general decline in educational levels and technical literacy of the general user population in the face of rapidly accelerating technology. As penetration of Third World markets increases, designers must also consider potential mismatches caused by cultural and educational differences as well. The mismatch problem has increased so much that it often serves as grist for comedians, who interview citizens on the street to demonstrate how little they know about the information that is part of daily life.

A national survey of Americans found that deficits in technological literacy were comparable to the decline in general literacy (Sum 1999). In surveys, investigators determined that:

- 50% of American industrial workers have math skills below the eighth-grade level
- 28% of workers cannot make correct change at cash registers
- 33% of people do not know how a telephone works

Too much knowledge or experience can be problematic as well. A phenomenon of *negative transfer* can occur when attributes of equipment, environs, and so on, suggest an action that is ultimately inappropriate. The user interacts with a product, piece of equipment, or tool based upon expectations created by past training and experience. Manual gearshift configurations, for example, are not universal in layout. Many times the configuration is placed on the gearshift knob, but the symbols are often worn off with use, and individuals experienced with one configuration will act as if using the configuration to which they are accustomed when introduced to different vehicles with unlabeled gear shifts.

Integrating human factors engineering requirements for any design should be determined at the outset—indeed, throughout the design development process. Integrating good human engineering design in the early stages of design reduces costs and the likelihood that post hoc modifications will have to be made, as well as safely increasing or restricting the breadth of user populations or of products, as desired.

Other problems develop when *quick fixes* are made to designs for any of a variety of reasons. Typically the original design team is involved in such design modifications and, without careful study, unintended problems can arise. It may be less expensive, for instance, to use a different-colored dye in polymers used to construct a product—so the color of the product is changed. But the color of the labels on the product may no longer offer the contrast sensitivity needed, causing operators to confuse controls, reducing functionality and potentially introducing safety hazards.

Sometimes designers forget to consider the capabilities and limitations of their intended and unintended user populations when making decisions about design parameters. The directions in which users will read symbols or icons can vary if a product will be used internationally. Some cultures read from right to left, others from top to bottom, still others from left to right, and so on. Differences in the viewing sequence of label information or graphics may alter the perception of the viewer. Failure to understand differences in the user population can make apparently safe designs suddenly unsafe.

Often designers forget that they are not merely designing a single, isolated product, piece of equipment, task method, or working or living environment. Rather, they are designing a component that will function within a larger system. Failing to consider this interaction often leads to poor design, suboptimal applications, or creation of hazards, all of which can go undetected. Often personnel requirements—the numbers and kinds of personnel who will be needed to staff the system, or those who can or are able to use the design—depend on the extent to which the systems approach is followed.

There are a variety of explanations for the development of HFE design flaws and safety hazards. Good human factors engineering design can only be achieved through rigorous analysis and comprehensive application of good design principles in the initial and the modification design phases. Project managers often misjudge the amount of effort required to engineer equipment, products, processes, and environments that must be functional and safe when they require interaction with humans. Nearly all engineering curricula do not require engineers or other professions to take coursework that integrates HFE design contributions into the design process.

Simply providing textbooks, design handbooks, and focused HFE standards for use by design teams is not always helpful. Design guidance is often too general and must be selected and tailored to meet projects' requirements. Human factors engineers typically participate with designers, helping them tailor their designs to meet users' needs, improve performance, and reduce the risk of causing safety hazards. Often, HFEs design tests and evaluation studies to confirm that design objectives have been met.

THE SCOPE OF HFE PRACTICES

The intent of this chapter is to illustrate the scope and usefulness of HFE methods in preventing or reducing safety problems. A general approach is recommended to address major problems that will likely be encountered. One cannot rely solely on recommendations, guidelines, checklists, or standards to eliminate HFE problems or hazards. Guidelines are often too broad or too specific within the context of a particular THEM system to provide material support. A systems approach to design and evaluation is proposed to manage the risk of producing unsafe designs.

No short chapter can fully address all the theoretical foundations, methodologies, and tactics used by HFE professionals to reduce safety hazards. No attempt has been made to comprehensively summarize a wealth of knowledge, design methodologies, or approaches to safe design using HFE, but many excellent sources of such information do exist in which readers can find detailed information concerning such methods—including many of the chapters within this book.

This chapter is intended to provide an overview of the field of HFE and its role in promoting safety in the workplace, at home, and in recreational environments. It focuses on human sensory, cognition, and motor-performance limits and their roles in designing and maintaining safe environments, providing a methodological approach that can be followed to achieve these objectives.

THE GENERAL HFE PROCESS

Human factors engineering professionals prefer to follow a standard sequence of operations when assisting in the design, development, and implementation of THEM systems. Human factors inputs are required throughout system development and generally follow the process below:

- Understand the objectives and goals of the system and their impact upon human roles and performance requirements.
- Provide information about human performance's capabilities of shaping of the system's design, about functional allocation between humans and machines, and about potential risks for system failure and safety and health problems.
- Evaluate THEM system performance.
- Discover whether human–machine system performance meets design criteria.
- Discover whether safety standards involving humans are met.

Designs in complex systems are fluid; the steps above can be followed iteratively and heuristically, capitalizing on previous designs, documentation, and test results. At each step of the process, judgment depends greatly on the skill and expertise of those who are conducting the analysis.

Although it is to be hoped that the above work could be performed by means of handbooks, standards, and design guidelines, it is likely that new ground will have to be broken and that such activities will require experimentation, tests, and evaluations of behaviors before final design recommendations are made or specifications solidified.

General Steps

Although specific expertise and distribution of HFE activities will vary from design to design, human factors engineers tend to follow a systematic process for design development, review, and validation. The general process followed is outlined below:

Step 1. Define specific system performance objectives and constraints

Without a clear understanding of the system's objectives and design constraints, it is difficult to discover which designs or design options are appropriate and which risks of faults, failures, or other problems are tolerable. Typically, outcome metrics from the test results judged are provided, as is other HFE information describing the intended user population, the desired staffing levels, the training and skills of the user population, the performance envelopes involved, and any other information useful to the design team.

If the performance objectives and constraints are not clear at the outset of the project, this usually spells trouble in meeting time and cost milestones. Such a project is typically not well thought out, and corners may be cut, increasing the risks of errors in design accordingly.

Performance specifications delineate the goals and performance requirements for a design or system. Specifications are typically prepared by a team that is involved in the development of the system. It itemizes functions, defines parameters, and spells out design constraints. The design team creates or responds to a "design to" specification that states system performance capabilities and user requirements.

The specification typically translates the user's operational needs into system functions and requirements, allocates those requirements to subsystems, and allocates general functions to operators and service personnel. Other than expected or required functional allocations, the specifications do not address specific requirements in terms of human performance beyond gross statements about personnel requirements (e.g., will be used by airline pilots, firefighters, the general public, and so on).

The objective at this stage of design is to determine how the system design specifications map onto human performance requirements or demands. Essentially, the HFE must determine what the performance envelopes are and feed this back to the design team—particularly if the requirements are unreasonable.

To meet this objective, system requirements must be project-specific and written in a verifiable form. If the requirements are too broad, the HFE and other team members will face great difficulty attempting to prove that they have met their goals. Establishing requirements in operationally defined, verifiable manners provides a common perspective and basis of understanding among the design team and its sponsors or customers.

Step 2. Allocate roles, responsibilities, and performance requirements

It is important to be apprised of the roles of humans in meeting the system design objectives and requirements. It must be understood how performance demands are to be functionally allocated among humans, machines, software, and other system elements. HFEs typically document the basis for task allocation, performance demands, and safety responsibilities among humans, firmware, and software.

Step 3. Perform actual or simulated task or activity analysis

Activity analyses may be performed using a variety of approaches. However, functional, decision, and action flow analyses are typically used, along with simulations and mock-up analyses. Examination of similar systems and focus groups are used as resources for understanding human performance and the need for allocated responsibilities. Job safety analyses can also be performed during task analysis to discover sources of failures and accidents, as well as and failure modes and effects.

Step 4. Define design questions or problems encountered

It is inevitable that design questions or problems will develop during Step 3. Questions will arise of which method is best to employ, of the safety of performing a particular task or of human interaction with machines or the working environment, or of the possibility of using new technology.

It is important to develop concise definitions of design questions or problems to define the scope of the effort of the HFE. Bounding the question or solution space enables the development of responsive answers in a timely, cost-efficient manner. Knowing when a functional answer to a designer's question has been produced allows efforts to be redirected to answering succeeding questions with confidence. It also makes validation of the recommendations more straightforward and efficient.

Step 5. Understand the design questions

Interrelating design questions with system performance requirements is critical. If these steps are decoupled, a design can be recommended that optimally addresses the design question within a particular subsystem, but it may not yield an optimal outcome for the system at large. Design recommendations for controls, displays, and seating design in a race car intended to travel at the vehicle's maximum capacity may be entirely inappropriate and overly costly, as well as promote inappropriate operation on highly trafficked downtown city streets.

Step 6. Select candidate design concepts for evaluation and testing

Often design options are provided by the designers. HFEs endeavor to discover which design candidate is best from a human–machine performance standpoint within the context of overarching system performance objectives. The HFE may also find design modifications that can improve any given candidate's design value.

Experimentation in laboratory or field conditions is usually required in order to test design options. Analysis, presentation, and interpretation of results are considered by the design team. Testing must be timely and cost efficient, but experimental findings should have adequate statistical force. Depending on the nature of the questions or problems addressed, testing and evaluation may iterate until a satisfactory design state is achieved by means of performance selection or utility metrics.

If multiple performance factors must be addressed, HFEs usually work with design teams to determine the weights to assign to each performance metric. The sum of the product of weighted or valued performance metrics is often used to rank design concepts by multiattribute utility (or by means of some form of decision support analysis).

HFEs should list all documents that will be or have been consulted in the development of the system. Any tailoring of information derived from standards should be documented. Ambiguously worded requirements in standards should be reworded or operationally defined so that HFEs are not later in disagreement with customers about whether a requirement has been met.

Step 7. Evaluate personnel selection and training requirements

If personnel selection criteria are set as part of the system design or performance requirements, analyses should confirm that the recommended design characteristics or candidate design options have not exceeded the intended user-population capacities. If that is not feasible, it must be determined what population selection criteria or training requirements are now requisite to meet the system's mission or performance requirements. Failure to carefully evaluate personnel selection and training needs, or to document them for future use, inevitably creates problems that promote accidents.

Step 8. Document and justify design recommendations

Documentation of standards relied upon, rationale for tailoring standards to address design questions, testing methods, data analyses, findings and interpretations of findings, and rationale for recommendations must be documented thoroughly.

Computer programs, drawings, mockups, and detailed reports should be delivered for use and archiving. The number and structure of reports, as well as the detail required, can vary from project to project. However, in the author's experience, most HFE reports include:

- failure modes and effects analyses (FMEA)
- management oversight and risk tests
- human factors system function and operator task analysis
- human factors design approach
- human factors test or simulation plans
- procedural data documentation and analysis reports
- personnel and training plans

Other Considerations

Training Team Design Members

Short courses addressing the general utility of HFE are useful, but an intense course over a couple of days or hours cannot be expected to provide designers with the capacity to apply HFE principles in the design of complex human–machine systems. Although designers will become sensitive to general concepts, they will not be able to evaluate engineering tradeoffs or address specific questions.

That said, sensitivity training for team members across disciplines is always beneficial, because it serves to enhance collaborative interaction and appreciation for problems in their colleagues' domains of expertise. Of course, HFEs must be prepared to provide design recommendations that are well-defended and tutorial in nature. Designers are not happy about altering their preliminary, intended, or already realized designs without solid justification. Constructing well-documented, thorough justifications for improvements in designs helps HFEs both to gain acceptance and to provide on-the-job training for design teams. Helping design teams learn why design principles are applicable often helps integrate such principles into future preliminary designs, expediting HFE review and (it is to be hoped) reducing the need for future design changes.

Understanding the Level of Effort

Documentation is requisite in HFE design and justification, particularly when systems under development are large and may require years to produce, and when the need for future modifications or updates is anticipated in response to expected changes in system requirements, available technology, or user populations. All analyses, preliminary studies, tests and evaluations, and tailoring or shaping of standards in their application for the project must be documented. It is not unusual for design and engineering team members to move in and out of the project, thus requiring documentation review by onboard team members or project managers.

Documentation is also requisite for investigators of accidents, whether in defense of torts or product liability disputes or in cases of future designs that may need to build upon existing ones. If project performance objectives change, previous documentation can aid in understanding the implications of making such changes, as well as areas in which additional testing and evaluation will be required.

However, methodical analysis, documentation, and communication with design team members consume vast amounts of time. Because of the initial cost of HFE contributions, it must be ensured that outcomes make the initial investment and subsequent modifications cost effective; starting over with an HFE evaluation of complex systems is untenable.

DETERMINING HUMAN INTERFACE REQUIREMENTS

Once the HFE and team understand the functional flow of the work performed within or by the system, an even greater understanding of the actual tasks performed by humans is required, as well as how they interact with tools, equipment, other humans, their environment, and so on. Such information is requisite to beginning the mapping of functional demands on human perceptual, cognitive, and motor-performance requirements.

Human activity analysis may include a variety of analytical tools that help define human performance requirements for an existing or future system design (Chapanis 1965):

- function allocation
- function and operational flow analysis
- decision action analysis
- action information analysis
- task analysis
- timeline analysis
- link analysis
- simulation
- controlled experimentation
- workload assessment

Functional Flow Analysis

Functional flow analysis is a procedure for decomposing a system design into functional elements and identifying the sequences of functions or actions that must be performed by a system (Chapanis 1965). Starting with system objectives, functions are identified and described iteratively; higher, top-level functions are progressively expanded into lower levels containing more and more detailed information. As demonstrated in Figure 2, functions or actions are individually and hierarchically numbered in a way that clarifies their relationship to one another, permitting the tracing of functions throughout the entire system.

The spatial and numeric organization is structured so that anyone can easily trace the input's flow through functions to output. Some HFEs identify the function's output with a connecting arrow to the next function, adding additional inputs to the next function.

The flow is from top to bottom, then left to right on the diagram. Connector blocks are used to link functional flows from one page to the next; arrows enter function blocks from the left and exit from the right. Functions subordinate to the general function are placed below it, showing the normal sequence of system functions—top to bottom, then left to right, and, if necessary, down. Whenever arrows join or split,

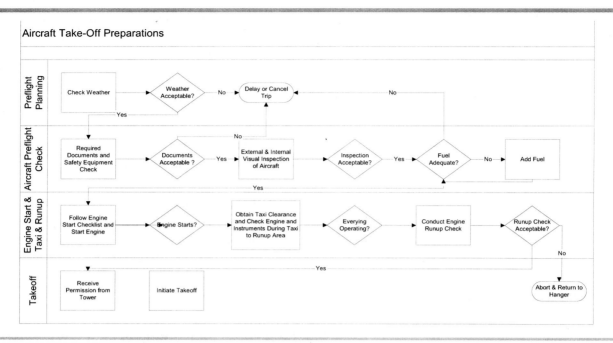

FIGURE 2. Example of a functional flow diagram

their junctions are shown, with "and," "or," and "and/or" gates encircled. An "and" gate requires all following or preceding functions to be or to have been performed; an "or" gate allows for one or more of the following or preceding functions to be or to have been performed.

For each of the functions, the team can indicate the need for decisions by posing binary questions. Each function is a short verb–noun combination incorporating occasional adjectives or other modifiers. One decision might be "Projector status OK?" This can be decomposed into lower-level decisions that include "Power On?," "Projection Mode Properly Set?," and "Projector Connected to Computer?".

Each decision is placed in a diamond symbol and is phrased as a question that may be answered with a binary (yes or no) response. Function blocks and decision diamonds are given reference numbers.

Operational Analysis

An *operational analysis* is an analysis of forecast functions in an undeveloped system and is designed to obtain information about situations or events that will confront operators and maintainers (Kurke 1961, Niebel & Freivalds 2003). Scenarios and anticipated operations are created, and assumptions and operating environments are documented. Scenarios should be sufficiently detailed to convey an understanding of all anticipated operations and should cover all essential system functions, such as failures (both hard and soft) and emergencies.

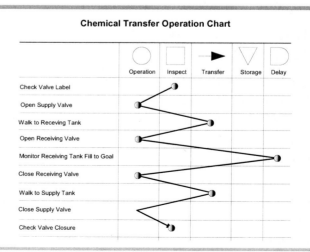

FIGURE 3. Operational analysis diagram and symbology

Operational analyses are typically driven by the system performance objectives and requirements. Scenarios are usually developed by knowledgeable experts that include procedures, equipment use, environmental constraints, and potential rare or unanticipated outcomes. The scenarios are delineated in detail sufficient to allow designers and analysts to work together on design formulation and assessment.

From an HFE standpoint, the operational analysis provides an interactive description of human constraints and requirements (e.g., environmental conditions, skill and training requirements, and so forth), performance envelopes that can affect system performance, and failure modes and effects that must be considered when specifying HFE design requirements. It is therefore important the team producing the operational analysis is experienced with the technology and the operations addressed, and that the content experts (including HFEs) be participants. Typical operational and sequence diagrams and symbols used are shown in Figure 3.

Task Analysis

Task analysis can focus on what actually happens in the workplace now but can also be used as a tool to specify the design of a new system. *Task analysis* provides a detailed description of how goals are accomplished, allowing designers to provide the means for enhancing human performance. Analysis involves a systematic breakdown of a function into its underlying tasks, and then of tasks into subtasks or elements. The process creates a detailed task description of both activities, both manual and mental; durations of tasks and elements; frequencies, allocations, and complexities of tasks; environmental conditions; necessary clothing and equipment; and all other unique factors required for one or more humans to perform a given task. Estimates are made of the time and effort required to perform tasks, as well as of perceptual, cognitive, and motor-performance demands; training requirements; and all other information needed to support other system-development activities.

Task analysis is the first step in taking a systematic approach to HFE design. Task analysis can focus on what actually happens in the workplace now or can be used as a tool to specify the design of a future system.

To decide what type of information should be collected—and how it should be gathered—it is necessary to identify the focus of the analysis. It should be decided not only what system is to be focused upon but also for what the results will be used (perhaps redesigning a new system, modifying an existing system, or developing training). Not all task information need be collected, but merely sufficient information to allow specific necessary objectives to be addressed (see Figures 4 and 5).

Task Information	Description
Task and Subtask Structure	An organized, often hierarchical listing of the activities involved in a task with tasks and subtasks numbered accordingly.
Importance or priorities of subtasks	Assessment of the criticality of subtasks from a performance or safety perspective.
Frequency of subtasks	Frequency of occurrence of subtasks under different conditions.
Sequencing of subtasks	Order of occurrence of subtasks under different conditions.
Decisions required	Part of the sequencing may be based on a decision needed to choose the branch of activity and thus a given set of subtasks.
'Trigger' conditions for subtask execution	Execution of a subtask may depend upon the occurrence of a particular event or a decision made during a previous task or subtask.
Objectives	Performance objectives are provided including the importance to the overall mission.
Performance criteria	Human performance criteria are delineated such as perceptual requirements, psychomotor performance, accuracy constraints, etc.
Information required	Information that must be provided at each level of task or subtask that must be attended or used by the operator.
Outputs	Information, products or other results of human effort that result from the subtask.
Knowledge and skills required	Information and skills that the user utilizes in decision making and task performance.

FIGURE 4. Information that should be gathered when performing task analyses (*Source:* DOD 1987)

Data Collection Method	Description
Direct and Videotape Observations	Use of time study or work sampling methods to develop job descriptions and psychomotor demands.
Interviews and Questionnaires	Interview workers or operators to determine the sequence of operations, tasks, subtasks, and hazards and difficulties they encounter in the performance of their job or task.
Focus group	Discussion with a group of typically 8 to 12 people, away from work site. A moderator is used to focus the discussion on a series of topics or issues. Useful for collecting exploratory or preliminary information that can be used to determine the questions needed for a subsequent structured survey or interview.
Interface Surveys	A group of methods used for task and interface design to identify specific human factors problems or deficiencies, such as labeling of controls and displays. These methods require an analyst to systematically conduct an evaluation of the operator–machine interface and record specific features. Examples of these methods include control/display analysis, labeling surveys, and coding consistency surveys.
Existing Documentation	Review any existing standard data sets from production engineering departments or other industries, operating manuals, training manuals, safety reports, and previous task analyses.
Link Analysis	A method for arranging the physical layout of instrument panels, control panels, workstations, or work areas to meet certain objectives, for example, to reduce total amount of movement or increase accessibility. The primary inputs required for a link analysis are data from activity analyses and task analyses and observations of functional or simulated systems.
Work Sampling Analysis	A method for measuring and quantifying how operators spend their time. Random or uniform sampling of activities are then aggregated over some appropriate time period (for example, a day), and activity-frequency tables or graphs are constructed, showing the percentage of time spent in various activities.
Checklists	Use a structured checklist to identify particular components or issues associated with the job. Available for a range of ergonomic issues, including workplace concerns, human–machine interfaces, environmental concerns.
Job Safety Analysis (JSA)	Using JSA, one can identify what behaviors in an operation are safe and correct. This analysis can be performed during task analysis. For each task or subtask, the analyst determines how the task should be performed and potential unsafe or hazardous methods. See chapters providing greater detail on JSA methods and benefits.
Critical Incident Analysis	Critical incidents can lead to accidents or system failures. This information is typically gained from human operators who supply first-hand accounts of critical incidents which are accidents, near-accidents, mistakes, and near-mistakes they have made when carrying out some operation. The critical incident technique only identifies problems—not solutions. The percentages of errors found in a critical incident study do not necessarily reflect their true proportions in operational situations, because the incidents are dependent on human memory and some incidents may be more impressive, or more likely to be remembered, than others.
	To be useful, the incidents must be detailed enough (a) to allow the investigator to make inferences and predictions about the behavior of the person involved, and (b) leave little doubt about the consequences of the behavior and the effects of the incident.
	The HFE then groups incidents into categories that have operational relevance:
	Mistakes and near-mistakes in reading an indicator,
	Mistakes and near-mistakes in using a control,
	Mistakes and near-mistakes in interpreting a label.
	The analyst then uses human-factors knowledge and experience to hypothesize sources of difficulty and how each one could be further studied, attacked, or redesigned to eliminate it. Studies are usually necessary to find ways of mitigating or eliminating those problems where elimination cannot be achieved based upon first principles alone.
Fault Tree Analysis	Some analysts use fault tree analysis in conjunction with critical-incident task analysis or job safety analysis or failure modes and effects analyses. Event trees, starting from a problem root such as equipment or other form of failure, follow through a path of subsequent system events to a series of final outcomes. This is also a probabalistic analysis that gives rise to likely failures and consequences that should be considered in the design of equipment, layouts, human capability demands, and so forth.
Failure Modes and Effects Analysis (FMEA)	FMEA is a method used to understand the consequences resulting from a failure within a system. Humans can be considered components that can fail for various reasons, and following FMEA methods, outlined in other chapters of this text, one can examine opportunities for eliminating or softening failure impacts. Typically, FMEA is used during functional flow, task, workload, linkage, and other HFE analyses. It helps in HFE analyses to determine what scenarios need to be interrupted by increasing the reliability of human–machine system performance.

FIGURE 5. Potential methods for data collection required for task analysis (*Source:* DOD 1987)

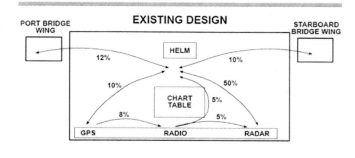

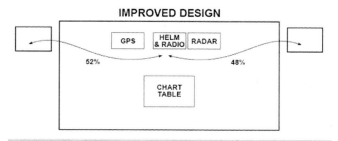

FIGURE 6. Linkage analysis showing relationships among ship bridge equipment placement and movement of an individual conning a vessel

Just like functional diagrams do, task analyses use hierarchical number systems to help analysts understand the relationships of subtasks, tasks, and jobs. Once a task or subtask is numbered, it bears that number for the remainder of the design project.

Linking Tasks Together

Link analysis is a method for arranging the physical layout of instrument panels, control panels, workstations, or work areas to meet certain objectives (such as reducing the total amount of movement and increasing accessibility) (Chapanis 1965, Sanders & McCormick 1993). A *link* is any useful connection between a person and a machine or part, between two persons, or between two machines or machine parts. If, for example, an operator walks to a supervisor to obtain permission or to consult about a problem, that activity represents a person–person link. If the operator returns to actuate the machine tool's cycle lever to press a metal part, that is a person–machine link. When the pressed part is automatically transferred to a transfer conveyor, that event represents a machine–machine link.

A number of steps are involved in link analysis:

- List all personnel and items to be linked.
- Determine frequencies of linkage among operators, tools, machines, and so forth.
- Classify the importance of each link.
- Compute frequency-importance values for each link.
- Sort link frequency-importance product values in order to focus on the links having the greatest cost.
- Fit the layout into the allocated space while minimizing linkage values.
- Evaluate the new layout in light of the original objectives.

Evaluation of various workplace or equipment arrangements can be made to assess move distances traversed during typical or unusual operations, as well as the crowding of activities and opportunities (clustering equipment to expedite operation) and the completion of multiple adjustments of areas, which then require inspection. Linkage analysis also elucidates the focus and concentration of human–human and human–hardware interactions, information that is very useful in weighing the importance or directing focus on design problems. See Figure 6 for an example of linkage analysis that shows relationships between ship bridge equipment placement and movement of an individual conning a vessel with a poorly laid-out bridge as opposed to a design that improves the operability of the vessel.

Workload Assessment

General workload assessment can also be performed at this point, using timeline analysis (Chapanis 1965, Sanders & McCormick 1993). Timeline analysis follows naturally from task and linkage analyses and is concerned with the scheduling and loading of activities upon individual operators. Charts are produced that show sequences of operator actions, as well as the times required for each action and the times at which they should occur. The method produces plots of the temporal relationships among tasks, the durations of individual tasks, and the times at which each task should be performed.

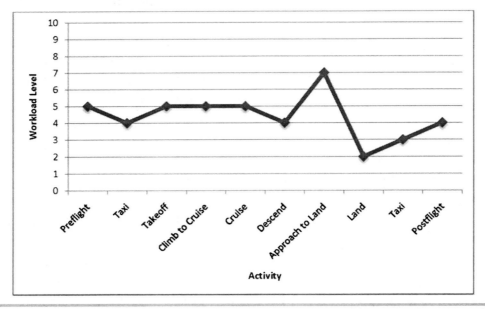

FIGURE 7. An example of workload timeline analysis used to identify activities where mental workloads could be high or low

Figure 7 shows that it is possible to integrate times from separate task analyses to show overlapping activities of two or more persons (Wickens & Hollands 2000). The greatest estimated workload occurs during the approach to land, and the least during landing.

Functional Allocation

Results of task analyses and workload timelines are often considered when making functional allocation decisions among workers, workers and machines, and among machines. *Functional allocation* is a procedure for assigning each system function, action, and decision to mixtures of hardware, software, and human operators. Allocation decisions can follow many approaches. They can range from *Fitts' Lists* of machine versus human performance strengths to *line-balancing* algorithms, in addition to results from experts, workers, and other inputs (Fitts 1951). For example, Fitts advocated the use of machines for massive counting or arithmetical tasks, pointing out that humans are better suited to handle novel events requiring complex assessment and decision making.

Professional judgment is involved in selecting criteria to be considered and in determining the weights to assign allocation criteria. In general, nearly all allocation decisions will have to address at least the following decision criteria (Price & Tabachnick 1968, Price 1985):

- implementation constraints
- maintainability and support logistics
- number of personnel required
- performance capability

- personnel selection and training costs
- political considerations
- power requirements
- predicted reliability
- safety
- technological feasibility

Analysis of Similar Systems

If possible, consider designs and analyses of similar systems, something that can increase understanding of important design issues that have been considered in the past (and which may be quite applicable to current problems) (Chapanis 1965). If access to reports, tests, and evaluations is possible, as well as to interviews of users of such designs and records of accidents or accident investigation reports, the work of previous design teams can be capitalized upon.

But HFEs should be careful to not merely blindly accept previous designs without considering the differences between current design requirements and those of previous projects. Antecedent systems often provide better initial insights into the operation and maintenance of proposed designs, into the skills and training required, and into any history of usability or safety problems, but a design is intended for use in different environments, by expanded user populations, or under more stressful operating environments; previous designs may not be adequate or may require modification.

Preliminary Focus Groups

Focus groups can help in assessing user needs, perspectives, and concerns early in the design process (Greenwood & Parsons 2000). A focus group's membership should be representative of the intended user group and small enough to encourage equal opportunity for expression of opinions and insights (e.g., five to ten members) about design issues and user needs. Preliminary focus groups can be much less formal than subsequent group meetings, in which greater moderation, and more focal or structured query will be needed to address specific issues.

MAP PERFORMANCE REQUIREMENTS ONTO A HUMAN USER

For each system function assigned to a human, the HFE works bottom-up in specifying perceptual, cognitive, and motor demands that must be met to achieve required task performance using specific displays, controls, workplace configurations, and all other design variables affecting operator performance and safety. Often, stipulating design requirements requires design tradeoffs. HFEs must convey to design teams where tradeoffs are likely to occur, as well as the consequences of such tradeoffs (Chapanis 1965, Sanders & McCormick 1993).

Simply citing specific standards or sections of standards will not convince designers to read and fully understand HFE guidance. Dumping a stack of textbooks, design handbooks, or relevant standards on a design team's table while expecting its members to absorb and internalize the information in a fast-paced design process is very unrealistic and, furthermore, invites errors. Instead, the HFE must mete out information as needed, clearly describing tradeoffs and their consequences or conducting preliminary analyses to help direct designs along successful paths. HFEs should provide ranges of specifications, describing consequences clearly enough for design teams to understand and making them easy to take into account in design decisions.

It is also a mistake to allow designers to try to use disparate standards or design handbooks, cherry-picking what they believe to be important or relevant. When inexperienced designers are confronted with only general recommendations, such information can be interpreted in different ways, some of which may not result in easily usable systems.

Some demands or requirements placed upon human users do not have well-defined answers or specifications. This can be frustrating both to HFEs, who are under time constraints to arrive at recommendations for design teams, and to design teams, who are in holding patterns as they wait for input from HFEs. Sometimes design teams cannot wait for the development of specifications and must move forward. This can pose problems if no opportunity will arise later to correct decisions.

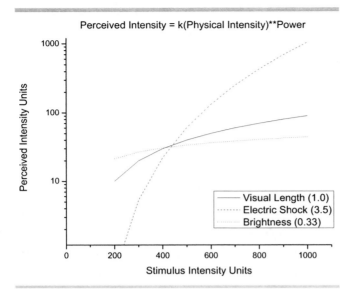

FIGURE 8. Impact of stimulus mode upon perceived stimulus intensity

TABLE 1

Various Stimulus Power Function Exponents for Several Stimuli

Stimulus	Power Exponent	Description
Brightness	0.50	Brief flash
Heaviness	1.45	Lifted weights
Lightness	1.20	Reflectance of gray papers
Loudness	0.67	Sound pressure of 3000 Hz tone
Muscle force	1.70	Static contractions
Pressure on palm	1.10	Static force on skin
Vibration	0.60	Amplitude of 250 Hz on finger
Vibration	0.95	Amplitude of 60 Hz on finger
Visual area	0.70	Projected square
Visual length	1.00	Projected line

(*Source:* Stevens 1957)

Are Acceptable Stimuli Presented to Users?

Is Stimulus Intensity Adequate?

A psychometric function describes the relationship between a parameter of a physical stimulus and the responses of a person who has to decide about a certain aspect of that stimulus (Stevens 1957). Change in perceived intensity with change in actual physical intensity obeys a power function.

Thus, different perceptual intensity changes are experienced with variations in actual stimulus intensity, and depending upon the stimulus type. See Figure 8 for the impact of stimulus mode on perceived stimulus intensity with the change in physical stimulus intensity based upon an exponent (noted in the legend).

Table 1 lists the stimulus power function exponents for several stimuli.

The psychometric function usually resembles a sigmoid function with the percentage of correct responses (or a similar value) displayed on the ordinate and the physical parameter on the abscissa. The presence of the stimulus parameter very far toward one end of its possible range indicates that an observer will always be able to respond correctly. Conversely, presence at the other end of the range indicates that the observer never perceives the stimulus properly, making the likelihood of correct response merely a matter of chance. In between these two extremes exists a transition range where subjects have an above-random rate of correct response but do not always respond correctly.

The inflection point of the sigmoid function, or the point at which the function reaches midway between random occurrence of correct behavior and consistently correct response, is usually taken as sensory threshold. A stimulus that is less intense than a human's sensory threshold will not elicit any sensation. Psychophysicists who study the relationship between physical stimuli and their subjective correlates, or percepts, have established thresholds for various stimuli that are used by HFEs.

Several different sensory thresholds are used by HFEs (Davis 2003; Fechner et al. 1966; Gescheider 1976; Gescheider 1984; Kaernbach, Schröger, & Müller 2004; Ljunggren et al. 1989; Manning & Rosenstock 1979; Psychonomic Society 1966; Stevens 1975; Yost, Popper, & Fay 1993).

Absolute threshold (AL) is the least stimulus that can be sensed 50 percent of the time by a (typically young adult) human focused on the source of the stimulus in the absence of other stimuli. Absolute thresholds are generally determined by plotting detection rates against stimulus intensities or characteristics.

Recognition threshold (RL) is the level at which a stimulus can be recognized 50 percent of the time.

Differential threshold (DL) is the level at which an increase in a detected stimulus can be perceived 50 percent of the time. This threshold is often referred to as the *just noticeable difference* (JND). Designers often want to use the detection of change in stimulus intensity as an indication of the state of performance of a machine, tool, product, alarm, or other such device. These thresholds are referred to as difference thresholds or difference limens (DLs) and refer to the amount of increase that is required in a sensed stimulus before one can detect a just noticeable difference. This magnitude depends upon the starting stimulus intensity.

Terminal threshold (TL) is the level beyond which a stimulus is no longer detected fifty percent of the time. The upper end of the detectable stimulus range is set by the inability of sense organs to respond to increased stimulus intensity (or to respond when such exposures are injurious). Some upper thresholds are set by consensus because of the difficulty of studying such exposures without injuring humans.

Thresholds, or *limens*, are set at the midpoint of a cumulative detection probability threshold, because the inflection point of the function provides the greatest resolution of sensory response to changes in the stimulus.

Equal increments of physical energy do not produce equal increments of sensation between lower- and upper-stimulus thresholds. Often designers anticipate that increments in stimulus intensities of signals can be arbitrarily used between lower and upper bounds of stimulus intensities with equivalent results, but that is usually not the case. The DL, at a particular stimulus intensity of a given stimulus, complies with an approximating fraction known as Weber's Law (or Weber's Fraction). A stimulus having a Weber Fraction of 0.10 need only be increased by 10 percent if a JND is to be detected. For a stimulus intensity of 10, the JND is 1. When the stimulus is increased to 100, the stimulus intensity must be changed by 10 units to produce a JND. The magnitude of change required depends approximately upon the stimuli's Weber Fraction.

For many sensory modalities, the JND is an increasing function of the base level of the current level of stimulus intensity (Van Cott & Warrick 1972; Mowbray & Gebhard 1958). The ratio of the JND, or ΔI, and the current stimulus intensity is roughly constant. Measured in physical units, we have:

$$\frac{\Delta I}{I} = k \qquad (1)$$

where I is the intensity of stimulation, ΔI is the change in stimulus intensity from the level I, and k is the ratio, fraction, or constant for that particular type of stimulus (often referred to as the Weber Fraction or Weber Constant).

TABLE 2

Relative Discrimination of Physical Intensities and Frequencies

Sensation	Number of JNDs
Brightness of white light	570 discriminable intensities of white light
Hues at medium intensities	120 discriminable wavelengths at medium intensities
Flicker frequencies between 1–45 Hz with on/off cycles of 0.5	375 discriminable interruption rates between 1 and 45 interruptions/sec at moderate intensities and an on/off cycle of 0.5
Loudness of 2000 Hz tones	325 discriminable intensities for pure tones of 2000 Hz
Pure 60 dBA tones between 20 and 20,000 Hz	1800
Interrupted white noise	460 discriminable interruption rates between 1 and 45 interruptions/sec at moderate intensities and an on/off cycle of 0.5
Vibration	15 discriminable amplitudes in chest region with broad contact vibrator within an amplitude range of 0.05–0.5 mm
Mechanical vibration	180 discriminable frequencies between 1 and 320 Hz

(*Source:* Van Cott & Warrick 1972, Mowbray & Gebhard 1958)

TABLE 3

Number of Absolutely Identifiable Sensations

Sensation	Number of Identifiable Levels
Brightness	3 to 5 with white light of 0.1-50 mL
Hues	12 or 13 wavelengths
Interrupted white light	5 or 6 rates
Loudness	3 to 5 with pure tones
Pure tones	4 or 5 tones
Vibration	3 to 5 amplitudes

(*Source:* Van Cott & Warrick 1972, Mowbray & Gebhard 1958)

Designers often overestimate detection percentage, forgetting that such values are only useful when humans are intensely attending to stimuli. A rule of thumb is that stimulus detection at 95 percent or 99 percent requires that the 50 percent detection threshold should be multiplied by 2 or 3, respectively. However, if an operator is not intensely focused on a stimulus, or if the stimulus is changing or moving, threshold multiples of 10 or greater may be required to achieve intended design objectives. Standards, handbooks, and testing may be required to arrive at stimulus intensities that meet design-requisite performance objectives.

Confusion in design can occur when designers wish users to detect changes in stimulus intensity level. Although a large number of JNDs exist for any given stimulus (see Tables 2 and 3), JNDs represent relative change-detection capabilities (most stimuli are not amenable to relative comparison) and can only produce 7 ± 2 absolute detection levels (Miller 1956).

If, for example, the intensity of a pure tone is presented as a cue or alarm for the severity of oil temperature, perhaps only two levels of sound power level should be used to cue the vehicle operator. Attempting to use many levels of any given stimulus intensity with intentions of providing absolute cues may not be successful, confusing users, who may then respond inappropriately (see Tables 2 and 3).

Many investigators have contributed to our understanding of the specific types and ranges of physical energy (typically referred to as stimuli) that fall within human perceptual capabilities. Much early work has since been catalogued by others (Mowbray & Gebhard 1958, Van Cott & Warrick 1972).

Today, many design guidelines and standards publish topically relevant thresholds for use by designers and engineers. Care must be taken, however, to control exposure to stimuli. If stimulus intensity is too great, users can experience sensory stimulus fatigue, causing psychometric function and threshold specifications to change dramatically.

Visual Acuity Problems

Of all the human senses, vision is likely most important to system design. Many issues must be considered when specifying the nature of visual imagery, but visual acuity, or target size, contrast, and magnitude of illumination, are questions that must be addressed.

Visual acuity refers to the ability to see spatial detail and recognize images. Historically, visual acuity is specified in minutes of arc, the angle subtended by the object viewed. For objects less than 10 degrees of arc, a small-angle tangent approximation can be used to estimate visual arc:

$$\text{Visual angle (min of arc)} = \frac{57.3 \times 60 \times L}{D} \quad (2)$$

where

L = the size of the object measured perpendicularly to the line of sight

D = the distance from the front of the eye to the object

Generally speaking, acuities can be grouped into four categories (Safir et al. 1976, Sanders & McCormick 1993):

- detection of presence
- vernier, or detection of misalignment
- separation, or detection of space between dots, parallel lines, or squares
- form, or identification of shapes or forms

All these forms of acuity are affected by a number of physical and physiological factors. Among the former are illumination, contrast, time of exposure, and color. Engineering design handbooks provide significant guidance on the impact of various combinations of these factors (Booher & Knovel 2003; Galer 1987; Hancock 1999; Salvendy 1987; Stevens 1951; Tufts College & U.S. Office of Naval Research 1951; Weimer 1993; Woodson 1981; Woodson, Tillman, & Tillman 1992).

Designers should remember that visual thresholds are set at 50 percent detection rates. The visual arc is normally increased by a factor of 3 to obtain a 99 percent recognition threshold, but the size of the target may have to be increased if illumination, contrast, or movement will inhibit detection. Users may also have subnormal vision. If so, inverting the Snellen ratio and multiplying the required visual acuity for normal vision can encourage recognition by those who have, say, 20:100 vision (the visual acuity value is multiplied by 100/20, or 5) if it can be reasonably expected that users may not, for example, have the benefit of eyeglasses when attempting to exit a burning building, or when personal protective equipment makes impossible the use of glasses or contact lenses.

Operator age can materially affect detection of contrast and color variance. Designers should understand that visual detection and recognition is dependent upon targets' contrasts and illumination. If illumination decreases at dusk inside buildings, or in vehicles used at night, the size of the target will have to be increased to provide an adequate level of contrast. Vibration and hypoxia (reduced oxygen content in the brain brought on by ascent to high altitudes or by the inhalation of certain kinds of gases) also reduce visual acuity. (Conversely, pilots who breathe pure oxygen at altitude at night frequently report a dramatic increase in the perceived quality or intensity of their color vision.)

If a designer relies only upon color as a differentiating code for stimuli or information, reduced illumination shifts vision from color to grayscale, making color cues difficult to discriminate. Color perception is also affected by the characteristics of the target (spectral content, luminance), by the environment in which it is viewed, and by the observer (Israelski 1978). Eight percent of men, and less than one percent of women, have some degree of color vision deficiency (Sanders & McCormick 1993).

The type of illumination used can materially affect color perception and discriminability. Two colors that appear identical in tungsten light may appear entirely different in daylight. High-pressure sodium lighting distorts almost all colors. Designers often fail to consider the extent to which variations in illumination characteristics can influence color detection across a wide range of operational environments.

Audition

Hearing is very important for human operators, because it allows communication by way of speech, as well as allowing users to attend to a variety of auditory cues such as bells, buzzers, beeps, horns, sirens, and so forth (Fisk & Rogers 1997, Salvendy 1987, Woodson 1981). Designers often rely upon auditory signals to alert multiple operators when visual stimuli cannot be relied upon to do so.

Like vision, *audition* is a complex sensory system that defies comprehensive description in this chapter. But it is less difficult to describe a process that most HFEs go through in assessing the acceptability of speech or auditory display design. The transmitted information should match generally accepted selection criteria for auditory transmissions (Sanders & McCormick 1993):

- simple and short
- calls for immediate action
- does not refer to previous transmissions
- deals with events in time
- alleviates operators' overloaded visual systems
- supplements operation in areas of poor illumination
- provides instruction when absence from visual displays is required
- acoustical (within range of operators' hearing)

Discrimination between frequencies and temporal sound patterns is much more reliable than is attempting to discriminate between simple increases in the sound intensity of a pure tone. To avoid ambiguity, systems should use no more than about six different auditory signals. Use of previously learned signals is preferred, and increases in repetition rate, rather than in amplitude, should be used to convey urgency (Woodson 1981; Woodson, Tillman, & Tillman 1992).

The chief challenge of the use of speech is in producing intelligible communication within a noise field. Noise in occupational environments, during emergencies, or during storms, among other situations, can mask speech that is not sufficiently loud. Normal speech averages about 65 dBA, and shouting can nearly reach 100 dBA. An extensive bibliography can be found elsewhere (Sanders & McCormick 1993).

By measuring the percentage of correct words transmitted from a speaker to a listener, a speech intelligibility score can be arrived at (e.g., percent correct). Studies have shown that speech intelligibility decays when the signal-to-noise ratio narrows across the voice spectrum.

When the one-third-octave band frequency *sound power level* (SPL) of a speaker's voice and that of the surrounding noise field is measured (the difference between the voice SPL and noise SPL, in dBA, is the signal-to-noise ratio of the speaker's voice) and the difference at each frequency band is weighed across the octave bands, the sum of those weighted differences is known as the *articulation index* (AI). The AI ranges between 0 to 1; 0 represents unintelligibility and 1 perfect speech intelligibility (Kryter 1985). See Table 2 for the frequency-band weightings and an example of a solved AI computation of one-third-octave band weights for voice-to-background noise signal-to-noise ratios.

The AI can then be used to look up operational consequences or capabilities for speech communication.

$$\text{AI} = \frac{W_i}{30\,\text{dB}} \sum_{i=1}^{5} (\text{dBA}_{\text{speech}} - \text{dBA}_{\text{noise}} + 12\,\text{dB}) \qquad (3)$$

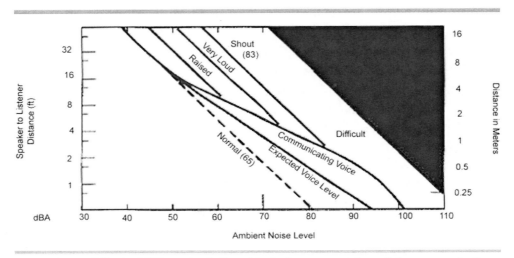

FIGURE 9. Effect of SIL and relationship between speaker–listener proximity and speech sound power level (*Source:* DOD MIL-STD 1472e)

where

W_i = weight for the ith frequency band

dBA_{Speech} = sound power level of speech at the ith frequency band

dBA_{Speech} = sound power level of noise at the ith frequency band

The average dBA readings for one-third-octave bands can be taken at 500, 1000, and 2000 Hz and their average used as the *speech interference level* (SIL) metric, querying the following figure to judge whether worker proximity and vocalization capacity are adequate for transmitting spoken messages in various noise fields (Sanders & McCormick 1993; Woodson 1981; Woodson, Tillman, & Tillman 1992). Figure 9 describes the effect of the SIL and the functional relationship between the proximity of speaker–listeners and the required speech sound power level.

Intelligibility may be enhanced by designing verbal messages that are simple and redundant, that use a reduced vocabulary set incorporating simple words, and that use phrases having contextual meaning (e.g., pilots expect to hear "cleared to land" on final approach).

Some designers assume that users of their systems will have adequate hearing, but hearing impairments can develop occupationally and with age. Unimpaired young operators can converse comfortably at 55 dBA. At the age of 50, speech sound power must average 67 dBA. At age 85, loud speech is required (about 85 dBA) (Coren 1994, Coren & Hakstian 1994). Because designers usually have no control over who will use their systems, most systems should be designed to allow for effective use by persons with hearing

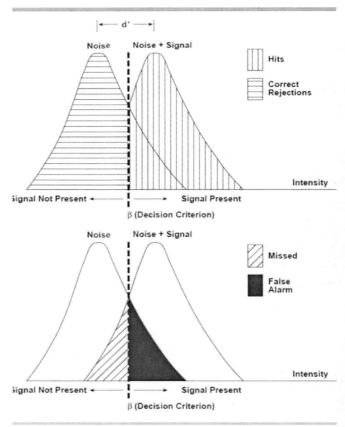

FIGURE 10. General stimulus signal detection behaviors

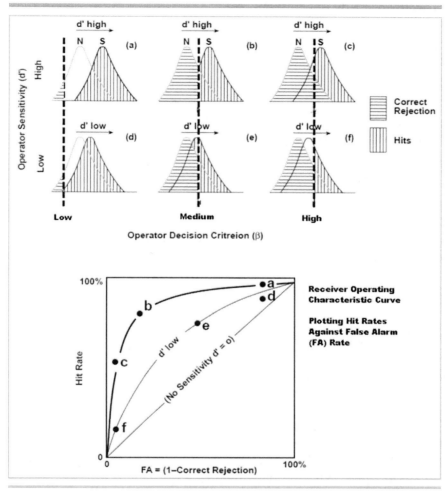

FIGURE 11. Impact of increasing β upon signal or hazard detection (Adapted from Wickens 2002)

difficulties. Providing volume controls so listeners can pick their own listening levels is important.

Are Signals Detected as Expected?

Engineers or designers often believe that if a stimulus is suprathreshold, it will be dependably detectable, but that is not actually the case (Green & Swets 1966, Green & Swets 1974, Hancock & Wintz 1966, Helstrom 1960, Macmillan & Creelman 2005, McNicol 1972, Poor 1994, Schonhoff & Giordano 2006, Swets 1996, Wickens 2002). As shown in Figure 10, observers can detect the presence (hits) and absence of suprathreshold stimuli (correct rejections), but they can also miss signals (misses) and report the presence of nonexistent stimuli (false alarms). The magnitude of each of these outcomes depends upon the observer's expectations regarding the presence of signals, upon their sensorial or perceptual sensitivity (d'), and upon their decision criterion (β).

The β used by an observer is the magnitude of the physical stimulus intensity at which point he or she will uniformly report the presence of a stimulus, as well as absence if the intensity drops. The observer's sensitivity, or inherent ability to detect the signal, can change with age, fatigue, and other factors that can degrade sensation. The observer's response criterion (β) can also change based upon the expectation of the presence of the signal and upon values and costs associated with judgments. If β changes in the face of a constant d', material differences will manifest in operators' accurate reports of signal presence and in their false-alarm rates (see Figures 10 and 11).

One can attempt to improve the observer's β by providing accurate information about the probability of signal and noise, the values of making a hit or correct rejection, and the positive magnitude of creating false alarms and misses. By adjusting the theoretical optimal β by a payoff matrix, observer SDT performance can be modulated:

$$\beta_{optimal} = \frac{PR(N)}{PR(S)} \cdot \left(\frac{Value\ (CR) + Cost\ (FA)}{Value\ (Hit) + Cost\ (Miss)}\right) \quad (4)$$

where

$\beta_{optimal}$ = optimal beta based upon payoff matrix costs and values

PR (N) = probability of noise

PR (S) = probability of signal

Value (CR) = absolute value of making a correct rejection

Value (Hit) = absolute value of making a hit or detecting the signal

Cost (FA) = cost of making a false alarm or claiming the presence of a signal when it is absent

Cost (Miss) = cost of missing the signal

As shown by Equation 4, and by Table 4, observers' response criterion can be easily biased by changing their perceptions of the prevalence of signals and of the values and costs of their decisions. The observed β for an observer is the ratio of the ordinal probabilities of signal and noise at the decision criterion. Thus, if the β is large, the criterion shifts to the right (i.e., physical stimuli must be substantial before observers will report the presence of signals). If the β is small, the criterion shifts to the left; small physical stimuli intensities will cause conclusions that signals are present.

Errors in signal detection can provoke unanticipated behaviors when cues to initiate behaviors are missed or evoke inappropriate behaviors when signals were not presented. Signal detection performance depends upon an observer's: (1) expectation of a signal, (2) response criterion (β), and (3) observational sensitivity to the stimulus (d').

One can compute the observed operator's sensitivity, or d', by estimating the distance between signal plus noise distribution (SN) and the noise distribution (N) in units of z-scores, which are obtained by determining the miss and false-alarm rates and finding the z-scores from the means of each distribution to the response criterion.

Another method for evaluating an observer's d' and β involves plotting the hit rate against the false-alarm rate for various signal detection trials in which different frequencies of signals or payoff matrices are used to shift βs. The result is a receiver's or observer's operating characteristic curve. As the perpendicular distance of the curve from the diagonal increases, the observer's signal-to-noise ratio, or capacity to detect the signal, increases. The slope of a tangent to the curve is indicative of the observer's beta.

One can use the magnitude of the area within a polygon created by receiver operating characteristic (ROC) points and the diagonal as a comparative metric of d'. As the area of the polygon increases, so does the d'.

A ROC curve allows the comparison of an observer's d' and β by plotting HIT rates against FA rates for various βs, which are changed by altering the probability of signals or the payoff matrix, or by allowing the observer to select different confidence levels for each decision regarding the presence of a signal or hazard. The ROC can be used to compare the d' values for different equipment designs among observers for personnel selection or to evaluate the impact of ambient

TABLE 4

Different Values for the Locus of Optimal Physical Intensity or Optimal β

Probability of Signal and Noise		Values		Costs		Optimal β
Noise	Signal	Hit	Correct Rejection	Miss	False Alarm	
0.99	0.01	100	100	100	100	99.00
0.99	0.01	100	100	100	100	99.00
0.99	0.01	100	100	100	100	99.00
0.99	0.01	100	100	100	100	99.00
0.99	0.01	1000	10	1000	10	0.99
0.99	0.01	1000	10	1000	10	0.99
0.99	0.01	10	1000	10	1000	9900.00
0.99	0.01	10	1000	10	1000	9900.00
0.01	0.99	100	100	100	100	0.01
0.01	0.99	100	100	100	100	0.01
0.01	0.99	100	100	100	100	0.01
0.01	0.99	100	100	100	100	0.01
0.01	0.99	1000	10	1000	10	0.00
0.01	0.99	1000	10	1000	10	0.00
0.01	0.99	10	1000	10	1000	1.01
0.01	0.99	10	1000	10	1000	1.01

and potentially masking phenomena (e.g., fog, rain, or darkness) upon the driver's capacity. In the ROC example, given observers A, B, and C have greater d', or sensitivities, than do F, E, and D, none of whom share the same β, or decision criterion. Some observers are very liberal, seeking to increase hit rates at the expense of increased false alarms (FA) (e.g., A and D), but others are very conservative (e.g., F and C), requiring greater signal intensities before calling a stimulus a signal. The conservative βs reduce both hits and false alarms.

The importance of understanding SDT is that it allows for the anticipation of significant differences in human detection of a signal or communication given differences that depend on the expectation of a signal and the values and costs associated with it. A designer should understand that a stimulus being suprathreshold does not guarantee that humans will accurately respond to signals or correctly reject their presence.

Is the Information Confusing the Operator or User?
Even operators who reliably receive stimuli can confuse those that are very similar in nature or perception. Ambient noise can also mask attributions and produce miscommunications. An in-depth discussion of information-theory approaches to the assessment of information sent and received is provided elsewhere (De Greene & Alluisi 1970).

If an operator's task requires the identification or classification of information, the sensory task has become perceptual task. The goal is to receive ($H_{received}$) all information that is sent (H_{sent}). If that occurs, the two circles in the Venn diagram shown will superimpose: all information is transmitted (H_t) without loss (H_{loss}) and with no noise (H_{noise}). Loss of information includes sent information that is lost and noise that is erroneously classified as information.

To discover the amount of information transmitted by labels, pictograms, shapes, and forms, calculate the information that was sent and received and that which collectively represented noise and loss. Information is computed in terms of bits. In the case of a single event i, the information sent by that stimulus can be determined with the formula provided in Figure 12.

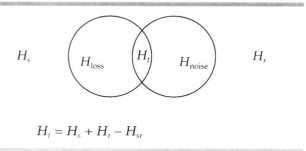

$H_t = H_s + H_r - H_{sr}$

FIGURE 12. General model for information transmittal (*Source:* De Greene & Alluisi 1970)

In the case of a large number of events, each possessing a different probability of occurrence, the average amount of information presented to the observer can be computed by the following formula for the average sent information:

$$H_{s\,average} = \sum_{i=1}^{N} P_i \log_2\left(\frac{1}{P_i}\right) \quad (5)$$

where

$H_{s\,average}$ = average information content (bits)
P_i = probability of event i

Suppose that, when four hazard icons, as shown in Figure 13, are presented to 1000 people, who are asked the perceived meaning of each of the icons, their responses are as shown in the data set in Figure 13.

The distribution of responses provided by the subjects showed that the skull and crossbones produces 900/4000 (or 23%) recognition as poison, and representing nearly half a bit of information:

By computing the average information sent (H_s = 2.0 bits), the information received from the marginal probabilities of the rows (H_r = 1.97 bits), and the information content inside the matrix (H_{sr} = 3.15 bits), the information that is transmitted can be determined (H_t = $H_s + H_r - H_{sr}$ = 0.81 bits). In the example above, one finds that the lost information ($H_s - H_t$ = 2.00 − 0.81 = 1.19 bits) caused by confusion and noise ($H_r - H_t$ = 1.97 − 0.81 = 1.16 bits) was equivalent and significant.

This analysis shows that there is hope for use of the skull and crossbones as an icon if one can discover why it was confused as a biohazard by 10 percent of the observers. Ideally, if one determines why the confusion between poisons and biohazards developed,

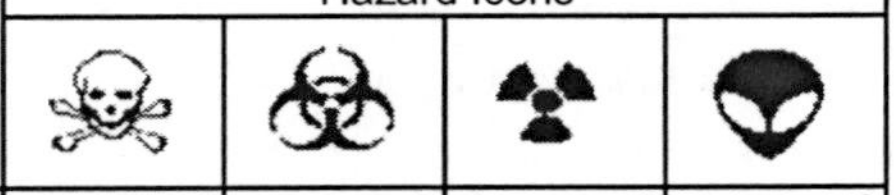

	Hazard Icons				Sums
	☠	☣	☢	👽	
Poison Hazard	900	0	0	0	900
Biohazard	100	0	300	300	700
Nuclear Hazard	0	500	400	300	1200
Alien Hazard	0	500	300	400	1200
Sums	1000	1000	1000	1000	**4000**

(Response on left axis)

	Hazard Icons				Sums	Bits
	A	B	C	D		
A	0.23	0.00	0.00	0.00	0.23	0.48
B	0.03	0.00	0.08	0.08	0.18	0.44
C	0.00	0.13	0.10	0.08	0.30	0.52
D	0.00	0.13	0.08	0.10	0.30	0.52
Sums	0.25	0.25	0.25	0.25	1.00	
Bits	0.50	0.50	0.50	0.50		

H_s = 2.00 Bits
H_r = 1.97 Bits
H_{sr} = 3.15 Bits

H_t = 0.81 Bits

FIGURE 13. An example of a confusion matrix and associated computations demonstrating the source and magnitude of symbol confusion in the user or test population

the design can be corrected. However, if the desired response to poisons and biohazards was identical, the confusion might be acceptable.

Often, the designers of icons, symbols, or other types of information that must be presented to users become too familiar with their designs and expect no confusion whatsoever. By conducting confusion matrix studies similar to the one in Figure 13 above, unexpected bases for confusion or equivocation among presented stimuli are often found.

Have the Affordances Been Adequately Considered?

One of the reasons that observers confuse the meanings of international symbols and other information is that affordances differ among user populations. Gibson described an affordance as an objective property of an object, or a feature of the immediate environment, that indicates how to interface with that object or feature (Gibson 1966). Norman refined Gibson's definition to refer to a perceived affordance: one in which objective characteristics of the object are combined with the actor's physical capabilities, goals, plans, values, beliefs, and interests (Norman 1988).

An *affordance* is a form of communication that conveys a percept such as an intended purpose, operation of a device, a message, or behaviors to be avoided. This is a powerful design element that can be useful if used wisely and can be punishing if it motivates inappropriate or undesired behaviors. Cognitive dissonance may also develop, leading to the use of objects in manners neither expected nor promoted by the affordance.

An example of a positive affordance is the steering wheel of a new car. A driver who has never driven the car before expects, based on the design of the steering wheel and previous experiences, that rotating the wheel will cause the car to veer in the same direction.

An example of an undesired affordance is a colorful, lightweight plastic gun that resembles a squirtgun but fires high-velocity projectiles that present significant risk of injury.

In sum, affordances can be very powerful influences in the receipt and interpretation of sent stimuli

Principles of Human Factors

that, if not evaluated through tests and evaluation procedures, can lead to dangerous behaviors.

Are the Cognitive Demands Acceptable?

Cognitive demands include handling information that is received from the sensory system, attended to, and combined with human memory to learn and to support decision making and selection of responses (Craik & Salthouse 2007, Durso & Nickerson 2007, Hancock 1999, Lamberts & Goldstone 2005).

HUMAN MEMORY LIMITS

Memory is an important tool for detection of problems, diagnostics, handling protocols, and learning new material, or learning from mistakes (Bower 1977, Cermak & Craik 1979, Estes 1975, Gardiner 1976, Shanks 1997). Memory failures often directly or indirectly cause human performance failures. There are generally three types of memory: *sensory, short term,* and *long term*. *Sensory memories* act as limited buffers for sensory input. Visual iconic sensory memory is briefly present for visual stimuli and aural stimuli, producing *echoic sensory memory*. Other sensory inputs have brief sensory-memory periods as well. Stimuli captured by sensory memory must move rapidly into short-term memory through attention. If the stimuli are not attended to, the sensory memory essentially filters that information out, and it fades away.

Short-term memory serves as a workbench for reinforcement of selected sensory memory and for temporary recall of information under process, and for assembly of more complex associations. Short-term memory decays rapidly if not sustained by continuous stimulus or rehearsal and pulls information from long-term memory to help develop associations, feeding associative structures into long-term storage.

Interference often causes disturbance in short-term memory retention, and rehearsal and refreshment of the working memory is necessary to minimize loss. Interference can be sensory, distracting of attention, or memory overload.

Long-term storage may be classified as *episodic memory* (storage of events and experiences in serial form) or as *semantic memory* (a record of associations, declarative information, mental models or concepts, and acquired motor skills). Information from short-term memory is selectively moved to long-term memory through rehearsal. *Rehearsal* of information, concepts, and motor activities enhances transfer into long-term memory. Learning is most effective if the rehearsal process is distributed across time.

Often, designers fail to: (a) provide sufficient opportunity for users to focus on sensory memory, (b) allow entry and rehearsal in short-term memory, and (c) enable users to establish associations allowing them to recall information from long-term storage. Moreover, designers can rely too heavily upon operator learning and recall to prevent errors in sequences of operations or to support diagnostic and predictive decisions.

Designing down memory demands is a good idea. Providing memory aids, such as checklists, increased display times, electronic to-do lists, attentional cues, large amounts of information *chunked* into smaller acronyms, and contextual cues can all promote accurate recall and sequencing of information. Effective recall is needed to make good decisions or to operate machines and tools and perform tasks properly, among other activities (Sanders & McCormick 1993).

Does the Design Support Human Decision Making?

Despite effort, humans are often irrational decision makers. Many times decisions are based upon past experience using heuristics or rules of thumb. In other cases, decisions can be flawed because of innate limitations in human decision-making capacities (Wickens & Hollands 2000).

A decision occurs when one must choose among options, predict or forecast outcomes, or diagnose a situation. The cognitive burden imposed by these activities is caused by having to recall or maintain a set of attributes or facts and their values while contemplating a decision. Metaphorically, one can view a complex decision as a long regression equation having many terms and coefficients. The greater the number of terms and coefficients, the more difficult the cognitive burden becomes, and the likelier it is that the decision will not be ideal. Choices are typically easier to make than are predictions, because predictions often require additional mental algebra (Wickens & Hollands 2000).

Diagnosis typically produces the greatest burden, because the individual must mentally regress from a current state into an array of possible etiologies. In the early stages of diagnoses of, for example, failure to start an engine, there may be many sources for such failure. Further data-gathering is typically required until the solution space can be adequately narrowed—but even when adequately narrowed, the potential solution space may be very large and can exceed human capacity to handle without high risk of error.

Attitudes or knowledge of factual attributes may take time to obtain and do not always arrive in an optimal or expected sequence or timeframe, which means decision makers may have to arrive at conclusions or make decisions when only partially informed. Initial attributes or weights associated with those attributes can decay with time, and decisions based upon an incomplete or erroneous attribute-weighting matrix can occur. Matching of weights to attributes may be out of phase, and early assessments based upon limited facts can lead to inappropriate initial hypotheses that, based upon prior experience or workload issues, are inappropriately adhered to. Initial hypotheses can serve as filters for new information as decision makers pay attention to only incoming facts that support initial guesses. Thus, from experience, decision makers attempt to arrive at decisions quickly to reduce the cognitive workload.

As the cognitive workload increases, as decision envelopes change, or as working memory is challenged, decision makers gravitate toward decision-making strategies that reduce their burden, an approach that often leads to error. Some of the types of errors encountered are summarized below.

When in doubt, humans often seek causality by correlation. When challenged, decision makers often seek causality by correlation. *Engine failures have a high correlation with exhaustion of fuel on long cross-country flights. Because an aircraft's engine stopped, it must have run out of fuel; an emergency landing should be attempted without further diagnosing the problem.* Leaping to such a conclusion may result in an unnecessary, dangerous emergency landing.

Rules of thumb or heuristic approaches to decision making simplify the process. Heuristics rely on a subset of attributes and exclude the need for further data-gathering. Intellectual bigotry rules out many possibilities, reducing the need to attend to certain facts or attributes, thereby expediting the decision with less burden. *Men generally have greater upper-extremity strength than women as a population. Because causing this accident required significant upper-extremity strength, the high-strength task had to have been initiated by a man.* However, population strength capabilities among men, women, and children overlap to such a degree that there may be insufficient evidence to exclude women from consideration in the accident investigation.

Humans are not objective, statistical, or computational machines. Statistical assessment of data usually takes place by intuition rather than calculation and leads to errors in assigning weights to attributes. Humans tend to linearize curvilinear relationships and thus over- or underestimate future system behaviors. Humans also tend to overestimate variability when means or ranges in sampled data are greater than comparison values, and they dislike arriving at estimates that are near extremes (i.e., they behave conservatively when estimating proportions). Estimates of means can be significantly influenced by modes, or the frequency of occurrence of a particular value (e.g., although the mean may be 20, there were six values of 4 and no other number was repeated, suggesting that the mean might be close to 4). Humans also do not reliably use Bayes' Rule. They do not accurately adjust future probabilities based upon prior knowledge of changes. Failure to correctly statistically characterize probabilities and magnitudes leads to a false perception of reality that can provoke inappropriate decisions.

When in doubt, humans often choose conservative decision outcomes. Essentially, they regress toward the mean. Humans bias their decisions based upon either primacy or recency factors. First impressions can take hold and bias all subsequent information-gathering and information-weighting. However, recent (perhaps negative) experience can shift the bias toward recent information, negating prior data or experiences. Either bias can provoke errors in assessment and decision. First impressions of workers are often incorrect, and a recent error that leads to a significant negative outcome (e.g., accident) does not necessarily reflect the worker's history of safe behavior.

Divide-and-conquer errors are often encountered in the face of overwhelming data and choices. Throwing too much information at a human often leads to filtering. Workers seek to accept a small set of hypotheses (usually less than 3 or 4) and attend to information that principally supports one or more of their initial guesses. Ease of recall of an initially feasible hypothesis may be used to filter additional information, and the worker may rely upon heuristics, primacy bias, and other biasing behaviors to control the mental workload of decision making.

Reliable or highly diagnostic sources can be overweighted by decision makers, causing inappropriate selection of hypotheses and faulty data-gathering. Experts used in accident investigations can provoke inaccurate conclusions because they have a demonstrated record of success and expertise. Second opinions and verification through testing, experimentation, simulation, or other objective measurements should be used to check expert opinions. One should also provide those opportunities when the expert is a machine, gauge, or computer display. Humans often believe that computers do not make mistakes, but in reality they often do.

Overconfidence in one's ability to make decisions. Past good performance as a decision maker can cause overconfidence, biasing perspective and expectancy and producing poor decisions.

Negative consequences outweigh positive consequences. Decision makers are likely to select outcomes that are risk aversive, avoiding costs rather than pursuing gains.

Confirmation and Negative Information Bias

Once humans believe they have the answer, they tend to develop *cognitive tunnel vision* and resist attending to information that contradicts their belief. Designers often believe that humans are objective decision makers. They expect humans to entertain all the facts presented, and to objectively weigh those facts and arrive at a mathematically consistent or optimal decision. Often this is found not to be the case, particularly when decisions are complex or multifaceted.

To help decision makers avoid poor decisions and negative consequences (such as accidents, injuries, or losses), designers should do the following (Wickens & Hollands 2000):

- Educate decision makers regarding what is causal and what is not.
- Reduce decision options as much as possible.
- Perform statistical computations and mathematical operations for decision makers and persistently present them for reinspection and reevaluation.
- Remind decision makers of all attributes that exist or that have been encountered, working to prevent their exclusion.
- Provide memory aids such as checklists and diagnostic fault trees to reduce the cognitive burden that makes cognitive shortcuts attractive.

CHECK MENTAL WORKLOADS

In preliminary analyses, it may be acceptable to use timeline analysis to estimate the mental workload, or timesharing, that an operator may experience. Essentially, try to balance out workloads among workers and machines in the system. However, once the preliminary design is developed, evaluate the extant design using experimental workload assessment methods to determine whether workers or operators are likely to experience excessive mental demands. Workloads may be acceptable during normal operations but excessive when failures occur or when operation occurs under stressful conditions (Wickens & Hollands 2000).

Mental workload measurement has been classified into three categories: *subjective* (i.e., self-reporting) *measures* (Gopher & Braune 1984), *performance measures*, and *physiological measures* (O'Donnell & Eggemeier 1986). Performance measures can be made of the primary or actual work task, of secondary operational tasks, or of non-operationally-relevant secondary tasks that tap the same resources that primary tasks do; these can be used to gauge workload by the degree of measured performance decrements.

The underlying rationale for workload measurement is that humans have many (but nevertheless limited) resources with which to perform work. When one or more underlying resources are overallocated, performance suffers and psychological and physiological stress develops. Absolute workload assessments can be made when directly measuring primary and secondary tasks that burden the same resource pool. Relative resource demand assessments can be made when comparatively evaluating indirect measures, such as physiological strain. Which type of workload measurement should be used depends upon a number of factors.

Wickens proposed a multiple-resource theory for workload assessment and control in which different resources exist for different modalities (Wickens 1984). Auditory and visual processing resources, central processing resources, and different motor resources are required for the performance of psychomotor tasks. Resources are classified by type and have dedicated channels. If work is designed to pass through one particular type of channel, the people involved may experience resource limitation problems, degraded performance, physiological strain, and other negative outcomes. Redesigning a task to keep inputs from sharing a signal channel and to allow outputs to use multiple channels can increase workloads (to a point) without running into resource allocation constraints. Wickens is careful to point out that workload design is a little more complicated than this, but the general principles appear to work well.

The multiple-resource model has stages and various capacity constraints. The processing stage addresses sensory encoding and perception, central processing, and motor-response planning. The auditory, visual, and tactile modalities draw upon different resources, allowing cross-modal timesharing to perform better than intramodal timesharing. Listening to the radio while tracking the position of a car in a lane is much easier than listening to the radio while listening with eyes closed to instructions about when and how much to turn the steering wheel to maintain lane position. Another facet of the model is the processing mode, which can be verbal or spatial in nature. And another facet of the model addresses input and response times. Overburdening a particular channel or switching between encoding, processing, and response modalities produces negative consequences for mental workload.

There are a wealth of studies and supplementary discussions on the nuances of human workload that cannot be adequately addressed here. However, the general concept of unbalanced or excessive use of resources creates problems in mental workload and thereby increases performance decrements, strain, and the likelihood that unsafe shortcuts will be taken.

Mental Workload Assessment

A number of metrics have been developed for assessment of mental workload. No single metric is ideal because they all vary in their sensitivity, diagnosticity, primary task intrusion, implementation requirements (e.g., availability of equipment, expense, and other logistical constraints), and operator acceptance (e.g., certain operator populations are very resistant to admitting any problems with their ability to handle task demands, fearing they will be bypassed for promotion; other operators may not honestly respond, putting honest responders at a disadvantage if the information is available to decision makers). Sensitivity is an index of the responsiveness of the metric to changes in workload. Diagnosticity is the ability to discern the type or cause of workload, or the ability to attribute it to an aspect, or aspects, of the operator's task (Wierwille & Eggemeier 1993).

Primary task intrusion occurs when measurement of the workload metric interferes with task performance, thereby giving false indications of workload problems. Secondary tasks can be used to assess remaining resource capacity after primary task requirements are made and the primary and secondary task share the same resources. Sometimes subjects switch primary and secondary task priorities when they know that only secondary task performance is being measured. Self-report measures taken after completion of the task, as well as most physiological measures, seem to degrade primary task performance the least (Eggemeier & Wilson 1991; Eggemeier, Wilson, Kramer & Damos 1991).

The reader should understand that substantial warnings applying to the use of mental workload metrics are given in the literature; a book would be required to adequately address them. No reliable decision algorithm exists for selecting one or more metrics beyond the selection criteria noted previously, the nature of the tasks performed, the nature of the operator population, and the nature of questions that have to be addressed in the assessment.

Errors have occurred in past measurement of mental workload, putting operators at risk. Often errors occur because designers have used inappropriate metrics (based upon the selection criteria noted above) or have not corroborated outcomes of a single metric (e.g., failed to use a battery of metrics). For example, one may measure primary performance and find no decrement, thus concluding that mental workloads are adequate. Yet such performance may have been achieved by extreme effort that produced psychological and physiological strain. With time, that situation will cause degradation of performance.

Another example is relying on a physiological metric merely because it is objective. Physiological metrics are integrated measurements that reflect the totality of exposure and often have lag times, offering little diagnostic value. Personality structure can influence stress responses; resource overloads can be missed in individuals who may not share concern about performance decrements.

Self-report tools offer great value but are subject to operator cooperation, which can be inhibited by poorly described procedures, neophyte operators, or managerial factors (e.g., if I report problems with this new system and others don't, will I be at a future disadvantage for selection to use it?). Designers who are strong advocates of their design may not be the individuals who should design, present, or assess the results of operators' self-reports via mental workload tools.

Dismissing weak indicators of mental workload problems or failing to consider the additional burdens that develop when components of systems are not operating as required or expected often produces situationally excessive workloads that go unaddressed during design. Often, HFEs are brought in after accidents and must reassess mental workloads to rule that problem out during an accident investigation. Mental workloads may be found to be high and inadequately considered during the design. More often modifications to equipment, jobs, environments, and so forth, result in increased mental workloads that go unassessed.

Complex systems are typically designed concurrently, as separate design teams work on different components of a design. Design teams may work too independently, failing to detect that HFE designs for a component system may not be appropriate when all components are merged and integrated into an operational system. Often operators' visual, auditory, and motor resources are overtaxed during certain performance scenarios that are not tested until the final stages of design. Handling system malfunctions or complex problems increases processing and motor burdens that exceed normally acceptable mental workloads. Mental workload assessments in component design may be useful at those levels. However, failing to replicate mental workload assessment at design completion or in the integration stages can be disastrous, particularly if systematic tests are not performed within an acceptable scope of failure modes.

ARE MOTOR-PERFORMANCE DEMANDS EXCESSIVE?

Motor response, or performance, can be classified as single, sequential, discrete open-loop movements, or closed-loop continuous movements. These activities can also be mediated by reaction-time components of manual performance. Textbooks recommended in this chapter's reference section have excellent, in-depth reviews of human reaction time and motor performance (e.g., Hancock 1999, Salvendy 1987, Weimer 1993). The following sections address sources of errors that designers have encountered when designing or evaluating the impact of their designs on motor-performance requirements and capabilities.

Are Anticipated Reaction Times Realistic?

The time taken by operators to detect and physically respond to some external event or input (such as a hazard signal or change in a traffic light) is referred to as *reaction time*. Reaction times are not static within operators; fatigue, attention or vigilance decrement, stress, aging, and other factors can degrade or enhance reaction time (Wiener 1964; Weiner 1987; Wiener, Curry & Faustina 1984; Wiener, Poock & Steele 1964). Reaction time is also influenced by the number of events or signals the operator must attend to. The Hick–Hyman Law demonstrates that increased numbers of events that must be handled, increased information content, and increased information-processing produce increased latencies in response times (Grossberg, AMA & SIAM 1981; Kamlet, Boisvert, & U.S. Army Human Engineering Laboratories 1969; Schulz & Lessing 1976; Welford & Brebner 1980):

$$RT = \frac{150 \text{ ms}}{\text{bit}} [\log_2(n + 1) \text{ bits}] \qquad (7)$$

where n = the number of choices or potential responses.

Reaction-time tasks are of three types: simple, disjunctive, or choice (Welford & Brebner 1980). Most reaction-time data have been collected under ideal laboratory conditions in which subjects are extremely focused on reaction-time tasks. As time-sharing increases, reaction times increase—sometimes demonstrably (Hancock & Caird 1993; Hancock et al. 1995; Hancock et al. 1990; Kramer, Trejo, & Humphrey 1995).

Thus, if a nuclear power plant operator must attend to 10 signal lights that indicate a malfunction, then responding by pressing a single master alarm button, the minimum amount of time needed to react to the signal can be determined. This choice reaction time is computed as:

$$RT = \frac{150 \text{ ms}}{\text{bit}} [\log_2(10 + 1) \text{ bits}] = 519 \text{ ms} \qquad (8)$$

Designers often err in performing restricted laboratory or field tests of reaction times, presuming that such times will continue to be reflected when the operator is exposed to many timesharing activities; but that will probably not be the case. In such situations, designers should carefully consider the consequences of long response latencies. If such outcomes are problematic, design interventions will be needed (e.g., computer-assisted vigilance, an engineer in soft failures, increased operator training and awareness, and so on).

Are Open-Loop Performance or Discrete Movement Times Realistic?

Open-loop tasks in which feedback is not or cannot be used, such as a discrete rapid movement of a hand to a control location, can usually be performed more quickly than can closed-loop or tracking tasks, in which feedback is used to adjust movements. Fitts' Law provides an open-loop motor-performance prediction model based upon the concept that different movement amplitudes to different endpoint or target accuracy constraints produce different information-processing demands for humans (Fitts 1954). Human information-processing rate capacity is limited, which means that one can anticipate that longer times will be required to perform longer movements or more precise movements, because more information must be processed within a channel capacity–constrained motor system.

Fitts and colleagues demonstrated a reliable relationship between movement times and task indexes of difficulty that were characterized in terms of bits of information (Fitts 1958, Fitts & Peterson 1964, Fitts et al. 1956):

$$MT = a + b \log_2(2A/W) \qquad (9)$$

where

a, b = regression model coefficients
A = amplitude or move distance
W = move endpoint accuracy requirement or target width

There are different degrees of molecularization of this model (Welford 1968; Wiker, Langolf, & Chaffin 1989). However, it serves as a warning to designers that if one designs movement tasks that approach the information-processing constraints of the motor system, errors and mental workload levels will develop to unacceptable levels. If the cognitive demands are already high, adding additional challenges (such as large amplitude movements to very precise endpoints, or very precise movements in general) forces trade-offs on the part of the operator. Thus, while the motor task may be handled, the operator will have to reduce timesharing to achieve such performance.

ARE MANUAL TRACKING OR CLOSED-LOOP PERFORMANCE DEMANDS EXCESSIVE?

Closed-loop tracking tasks occur when driving a vehicle, aiming at a target, or performing any task in which motor behavior is adjusted to control for error (see

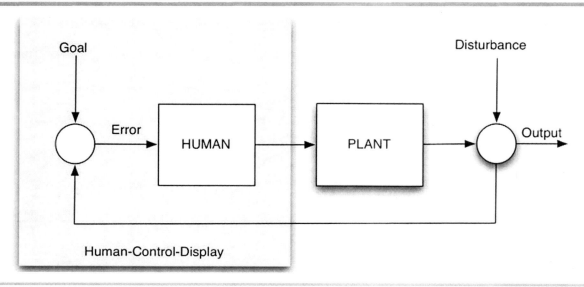

FIGURE 14. General model for human closed-loop tracking or control of a vehicle or other controllable device

Figure 14). The human attempts to govern the behavior of a physical entity that has inherent limits to respond. The physical system (such as an airplane with mass, damping, restoring, and other response phenomena) or plant responds to human control actions and environmental disturbances as the human operator attempts to reduce error in the airplane's response to control commands. Tracking can be pursuit (e.g., the operator sees both the current state or position and the goal or the target tracking feature, as when driving along a roadway or acquiring a moving target with a cursor or gunsight) or it can be compensatory (e.g., only the magnitude and direction of error is presented; the operator is asked to null the displayed error, as in an instrumental aircraft landing when glide-slope indicators show only horizontal and vertical deviations from the desired glide slope). Pursuit tracking is an easier task, because the operator can see and forecast movement of the target (Sanders & McCormick 1993).

Tracking performance is materially affected by control order. Zero-order controllers produce displacement outcomes in response to control displacement. First-order controls increase response velocity in correspondence to the magnitude of the control's displacement. Second-order controls increase response acceleration in correspondence to the magnitude of control displacement. Third-order controls relate the magnitude of jerk response to control displacement. In general, zero- and first-order controls produce best results and reduce mental workload demands. When higher-order controls are used, they are difficult to master and to use to achieve desired results (in most, but not all cases) (Sanders & McCormick 1993).

In addition to control order, designers often introduce response lags into systems, causing overcontrolling, phase errors, control-system instability, and other problems. Often poorly designed controls create control–display response incapability (e.g., tillers on sailboats versus steering wheels; the operator moves the tiller to the left to make the vessel turn right) that cause tracking mistakes, lost time, and accidents (Jagacinski & Flach 2003, Woodson 1981).

Controls often need their gains set. This is the magnitude of the response per unit change in control displacement. High-gain controls can be useful when large responses are needed to catch up with the target but are very problematic when the target is close and fine positioning (e.g., low-gain control) is required to acquire and maintain the target.

The operator's effective bandwidth in tracking tasks can be reduced by reducing control dead space (e.g., control slack or movement before underlying response is engaged) or control backlash (e.g., control

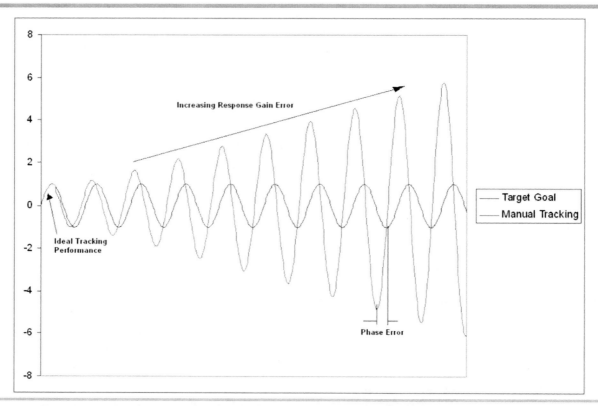

FIGURE 15. Examples of impact of phase and gain errors in human tracking performance (*Source:* Adapted from Jagacinski and Flach 2003)

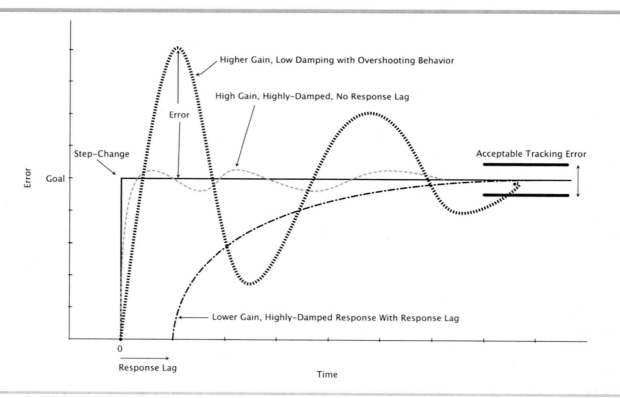

FIGURE 16. Examples of the impact of differences in control-system gain and damping characteristics upon human tracking response to a step-change in tracking goal or disturbance experiences by the human-machine system (*Source:* Adapted from Jagacinski and Flach 2003)

kick at the end of movement within the dead space). If the required bandwidth exceeds the operator's capacity, phase and gain errors can occur, causing loss of stability or control.

If the nature of the control design (e.g., gain, bandwidth, dead space, kick, momentum or inertia, compensatory versus pursuit tracking) is not matched well with the assigned task, operator performance can be significantly challenged, or fail. Matters only get worse when multiaxis or dual-limb tracking tasks are required. Figure 15 demonstrates the type of tracking errors that are typically encountered: excessive over- or under-shooting of the target goal, phase errors, or a combination of gain and phase errors.

Sometimes HFEs are brought in to resolve tracking task problems in which designers thought they were helping operators by introducing variable gains, variable control orders, and dual-tracking opportunities, often letting operators switch rapidly between modes. Whether or not such designs are helpful must be determined by realistic performance testing.

Control theory is typically used to model relationships between human–machine tracking requirements and tracking performance. Control engineers fit differential equations to tracking demands or forcing functions (e.g., step, random, or sinusoidal tracking requirements) and vehicle or equipment tracking response (i.e., tracking error). The relationship between the output and input behaviors is referred to as the *human–machine system transfer function*.

Typically, differential equations describing outputs, in response to known outputs, are converted to algebraic expressions using Laplace transforms. The simplified algebraic equation enables one to describe or predict tracking behvior for a range of input displacement magnitudes and rates of displacement changes. The resulting transfer function characterizes the lumped human–display–control–plant system behavior in terms of response, lag, gain, damping, and other performance-related characteristics. Knowledge of the transfer function can be used to predict the types of behaviors shown in Figure 16. Based upon this knowledge, one can "tune" the design of the control-display characteristics, as well as plant behaviors (e.g., aerodynamics of the aircraft). Details of this approach are addressed in Jagacinski and Flach (2003).

Designers often inherit control systems within the context of their new system design. Mental workloads and capacities to handle inherent tracking demands can differ materially from one system to another. Thus, looking at similar systems and planning on the control design that has been acceptable in the antecedent system does not guarantee that the design will be acceptable in a new application where perceptual, cognitive, and motor demands are quite different. Only by thorough testing in the context of system failures can one determine whether the control-system design is satisfactory.

ARE DESIRED OR SAFE BEHAVIORS MAINTAINED?

Behavioral-based safety refers to an approach that seeks to reduce unsafe behaviors through programs of behavior modification (Geller 1992, Kohn 1988, McAfee & Winn 1989, Peters 1989, Roberts et al. 2005). HFEs always attempt to design systems that do not require behavior modification or reinforcement. When that is not possible, personnel selection, training, and behavior modification and reinforcement are necessary—endeavors typically referred to as administrative controls.

Managers often think that short-term cost savings afforded by an administrative control are a better option than *engineering out* hazards. However, administrative controls must be maintained daily, and behavior must be managed; the energy and costs of such efforts can persist for long periods, eventually costing more than the engineering-control intervention. That said, HFEs often are faced with relying on administrative-hazard and unsafe-act control paradigms that are discussed at length in other chapters within this text. One example is the *NIOSH Work Practices Guide for Manual Lifting* (Waters et al. 1993).

SHOULD PERSONNEL BE SELECTED AND TRAINED?

Nearly all systems require some level of personnel selection and training (Bullard 1994, Bureau of Business Practice 1979, Kroehnert 2000, Leigh & Leigh 2006). The nature of personnel selection is determined by the perceptual, cognitive, and motor-performance demands that have been imposed by allocation of tasks within the human–machine system. Because other chapters deal with these requirements, this chapter will merely reinforce the need for HFEs to ensure that personnel selection factors are considered and carefully matched with the system's design requirements and behaviors. This is particularly important when attempting to accommodate potential users who are impaired situationally or chronically.

Training is often used to relax personnel selection criteria and pressures. The more complex and difficult the system is to operate, the greater the likelihood that it will need personnel who have higher levels of education, training, and skill. Training programs are likely to be more extensive as well.

Training programs should include operator understanding and recognition of hazards, safe behaviors, and failure-response or emergency protocols. After recognizing a danger, individuals must act to protect themselves against accidents and possible harm. They must know what actions are correct and safe and complete all actions required. Knowing appropriate actions and performing them correctly requires training, practice, and reinforcement through design and use of appropriate warnings, labels, and other hazard-response communications.

Even the best selection and training programs typically do not select personnel with the precise skills needed to operate new systems or to produce consistently desired responses. Training helps to reduce the risk of accidents and the impact or severity of accidents that do occur. Far too often, however, HFEs are confronted with heavy designer reliance upon personnel selection and training to handle design problems. Moreover, many accident investigations conclude that personnel selection and training must be improved to reduce risk of future accidents—code language for poor design that requires humans of greater capacities or who have more training and practice if operation is to have desired margins of safety.

DESIGN USING APPROPRIATE HFE STANDARDS, CODES, AND REGULATIONS

Standards, codes, and *guidelines* provide design information in a methodological manner (e.g., MIL-STD 1472D 1989; NASA-STD-3000 1995; ANSI/HFS 100 1988). Most standards are classified as specification standards where specific guidance is provided. The guidance may not yield the best design outcome, but it should yield, if applied properly, an acceptable design.

Performance standards do not provide cookbook-style guidance. Instead, performance standards help designers understand what the requisite performance criteria are for an area of application. The Occupational Safety and Health Act's general duty clause does not provide specific details, simply stating that any design that provides a job or workplace free from "recognized hazards that are causing or are likely to cause death or serious physical harm to . . . employees" is acceptable. The designer is free to seek a variety of design options that meet that standard of care. This usually requires demonstrable proof that the design is safe (e.g., testing or citation of other studies of like design that have proven to be safe in comparable tasks, equipment, environments, and worker populations).

Standards result from consensus among content experts from industrial firms, trade associations, academia, technical societies, labor organizations, consumer organizations, and government agencies. Failure to pay heed to standards and codes may be disastrous, particularly if a product, machine, or system is ever involved in a product liability lawsuit. Anyone who fails to consider standards, codes, or guidelines in professional safety is often asking for accident, injury, death, loss of property, and even appearance in court.

One may deviate from standards or guidelines, but justification for such deviation should be strong and should produce a better and safer design. Standards are consensus-based—in other words, reflective of lowest common denominators. Designers may need to improve performance or specifications to reflect a number of factors. However, although exceeding standards may need justification before the design team and those who authorize expenditures, substandard design is not acceptable.

It is often impossible for an evaluator or inspector, who is typically untrained in human factors, to decide whether a design does, or does not, meet a performance standard. A perfect world would see no opportunity for ambiguity or misinterpretation of HFE design standards, but that is simply not the case. Asking designers (who are not grounded in HFE engineering literature and principles) to interpret HFE standards without help is usually counterproductive.

Moreover, standards must be *tailored* or used with multifaceted design demands and constraints. Tradeoffs must be made to ensure that the intent of the standard is met. Sometimes standards present apparently contradictory or congruent information when, in fact, they do not. Moreover, not all requirements of a specification or performance standard must be *met* or designed to. Designers may spend significant amounts of time attempting to resolve apparently incongruent advice or requirements in a standard, a set of building codes, or a set of federal regulations. Terrible mistakes occur when designers latch onto specifications, use them, and remain unaware of an overriding specification in a separate standard to which they do not have access.

Standards, codes, and regulations cannot be used effectively by anyone who does not understand system performance objectives and requirements, as well as top-down design. Myopic use of the information provided in standards, design handbooks, building codes, and other consensus-based design guidance is an opportunity for misapplication of standards.

Design checklists do have their place, but far too often checklists are published without validation or assurance that, or even inquiry into whether, designs are effective and safe. Design checklists often act more as screening tools to grab the attention of those who have the education, skills, and expertise necessary to rule out problems by more thorough analysis.

Design checklists often query for binary responses on issues that are univariate in nature. Far too often, design problems represent interactions among a variety of design features. Additive responses may not provide any valid insight into the true magnitude of a problem or the acceptability of a design. Design checklists can be used to help detect the possibility of a problem, but they offer no protection when used blindly and improperly.

Conclusion

It is possible to design safe and effective systems that range from small- to large-scale, from simple to complex, and from development to testing. Human factors engineering principles work to improve the marriage between humans, machines, tasks, and operating environments. Good designs prevent accidents, mitigate the severity of the consequences of failures or accidents, and improve system performance. Those who are unfamiliar with HFE design principles should recognize the need to include HFE expertise in system design, testing, and evaluation from the outset, as well as when modifications are to be made to existing system designs.

References

Annett, J. 1971. *Task Analysis*. London: H.M.S.O.
Annett, J., and N. Stanton. 2000. *Task Analysis*. London: Taylor & Francis.
Booher, H. R., and Knovel. 2003. *Handbook of Human Systems Integration*. Hoboken, NJ: Wiley-Interscience.
Bower, G. H. 1977. *Human Memory: Basic Processes: Selected Reprints with New Commentaries, from The Psychology of Learning and Motivation*. New York: Academic Press.
Bullard, R. 1994. *The Occasional Trainer's Handbook*. Englewood Cliffs, NJ: Educational Technology Publications.
Bureau of Business Practice. 1979. *Training Director's Handbook*. Waterford, CT: The Bureau of Business Practice.
Bureau of Labor Statistics (BLS). 2007. *Census of Fatal Occupational Injuries*. Washington, D.C.: U.S. Department of Labor, Bureau of Labor Statistics.
Cermak, L. S., and F. I. M. Craik. 1979. *Levels of Processing in Human Memory*. Hillsdale, NJ: Lawrence Erlbaum Associates.
Chapanis, A. 1965. *Research Techniques in Human Engineering*. Baltimore: The Johns Hopkins University Press.
Coren, S. 1994. "Most Comfortable Listening Level as a Function of Age." *Ergonomics*. 37(7):1269–1274.
Coren, S., and A. R. Hakstian. 1994. "Predicting Speech Recognition Thresholds from Pure Tone Hearing Thresholds." *Percept Mot Skills* 79(2):1003–1008.
Craik, F. I. M., and T. A. Salthouse. 2007. *The Handbook of Aging and Cognition*. 3d ed. New York: Psychology Press.
Davis, S. F. 2003. *Handbook of Research Methods in Experimental Psychology*. Malden, MA: Blackwell Publishers.
De Greene, K. B., and E. A. Alluisi. 1970. *Systems Psychology*. New York: McGraw-Hill.
Department of Defense (DOD). 1996. MIL-STD 1472e, *Department of Defense Design Criteria Standard: Human Engineering*. Washington, D.C.: DOD

Desberg, P., and J. H. Taylor. 1986. *Essentials of Task Analysis*. Lanham: University Press of America.

Durso, F. T., and R. S. Nickerson. 2007. *Handbook of Applied Cognition*. 2d ed. Chichester: John Wiley.

Eggemeier, F. T., and G. F. Wilson. 1991. "Performance-Based and Subjective Assessment of Workload in Multi-task Environments," in *Multiple-Task Performance*. London: Taylor & Francis, pp. 217–278.

Eggemeier, F. T., G. F. Wilson, A. F. Kramer, and D. L. Damos. 1991. "Workload Assessment in Multi-task Environments," in *Multiple-Task Performance*. London: Taylor & Francis, pp. 207–216.

Estes, W. K. 1975. *Handbook of Learning and Cognitive Processes*. Hillsdale, NJ: Lawrence Erlbaum Associates.

Fechner, G. T., H. E. Adler, D. H. Howes, and E. G. Boring. 1966. *Elements of Psychophysics*. New York: Holt Rinehart and Winston.

Fisk, A. D., and W. A. Rogers. 1997. *Handbook of Human Factors and the Older Adult*. San Diego: Academic Press.

Fitts, P. M. 1951. "Human Engineering for an Effective Air-Navigation and Traffic-Control System." Columbus: Ohio State University Research Foundation.

_____. 1954. "The Information Capacity of the Human Motor System in Controlling the Amplitude of Movement." *J Exp Psychol* 47(6):381–391.

_____. 1958. "Engineering Psychology." *Annual Rev Psychol* 9:267–294.

Fitts, P. M., and J. R. Peterson. 1964. "Information Capacity of Discrete Motor Responses." *J Exp Psychol* 67:103–112.

Fitts, P. M., M. Weinstein, M. Rappaport, N. Anderson, and J. A. Leonard. 1956. "Stimulus Correlates of Visual Pattern Recognition: A Probability Approach." *J Exp Psychol* 51(1):1–11.

Galer, I. A. R. 1987. *Applied Ergonomics Handbook*. 2d ed. London: Butterworths.

Gardiner, J. M. 1976. *Readings in Human Memory*. London: Methuen.

Geller, E. S. 1992. *Applications of Behavior Analysis to Prevent Injuries from Vehicle Crashes*. Cambridge, MA: Cambridge Center for Behavioral Studies.

Gescheider, G. A. 1976. *Psychophysics: Method and Theory*. New York: Lawrence Erlbaum Associates

_____. 1984. *Psychophysics: Method, Theory, and Application*. 2d ed. Hillsdale, NJ: Lawrence Erlbaum Associates.

Gibson, J. J. 1966. *The Senses Considered as Perceptual Systems*. Boston: Houghton Mifflin.

Gopher, D., and R. Braune. 1984. "On the Psychophysics of Workload: Why Bother with Subjective Measures?" *Human Factors* 26:519–532.

Green, D. M., and J. A. Swets. 1966. *Signal Detection Theory and Psychophysics*. New York: John Wiley.

_____. 1974. *Signal Detection Theory and Psychophysics*. Huntington, NY: R. E. Krieger Publishing Company.

Greenwood, J., and M. Parsons. 2000. "A Guide to the Use of Focus Groups in Health Care Research: Part 2." *Contemp Nurse* 9(2):181–191.

Grossberg, S., American Mathematical Society (AMS), and Society for Industrial and Applied Mathematics (SIAM). 1981. *Mathematical Psychology and Psychophysiology*. Providence, RI: AMS.

Hancock, J. C., and P. A. Wintz. 1966. *Signal Detection Theory*. New York: McGraw-Hill.

Hancock, P. A. 1999. *Human Performance and Ergonomics*. San Diego: Academic Press.

Hancock, P. A., and J. K. Caird. 1993. "Experimental Evaluation of a Model of Mental Workload." *Human Factors* 35(3):413–429.

Hancock, P. A., G. Williams, C. M. Manning, and S. Miyake. 1995. "Influence of Task Demand Characteristics on Workload and Performance." *Int J Aviat Psychol* 5(1):63–86.

Hancock, P. A., G. Wulf, D. Thom, and P. Fassnacht. 1990. "Driver Workload During Differing Driving Maneuvers." *Accid Anal Prev* 22(3):281–290.

Helstrom, C. W. 1960. *Statistical Theory of Signal Detection*. New York: Pergamon Press.

Israelski, E. W. 1978. "Commonplace Human Factors Problems Experienced by the Colorblind—A Pilot Questionnaire Survey." Paper presented at the Human Factors Society 22nd Annual Meeting, Santa Monica, CA.

Jagacinski, R. J., and J. Flach. 2003. *Control Theory for Humans: Quantitative Approaches to Modeling Performance*. Mahwah, NJ: Lawrence Erlbaum Associates.

Jonassen, D. H., W. H. Hannum, and M. Tessmer. 1989. *Handbook of Task Analysis Procedures*. New York: Praeger.

Jonassen, D. H., M. Tessmer, and W. H. Hannum. 1999. *Task Analysis Methods for Instructional Design*. Mahwah, NJ: Lawrence Erlbaum Associates.

Kaernbach, C., E. Schröger, and H. Müller. 2004. *Psychophysics Beyond Sensation: Laws and Invariants of Human Cognition*. Mahwah, NJ: Lawrence Erlbaum Associates.

Kamlet, A. S., Boisvert, L. J., and U.S. Army Human Engineering Laboratories. 1969. *Reaction Time: A Bibliography with Abstracts*. Aberdeen Proving Ground: U.S. Army Human Engineering Laboratories.

Kirwan, B., and L. K. Ainsworth. 1992. *A Guide to Task Analysis*. London: Taylor & Francis.

Kohn, J. P. 1988. *Behavioral Engineering Through Safety Training: The B.E.S.T. Approach*. Springfield, IL: C. C. Thomas.

Kramer, A. F., L. J. Trejo, and D. Humphrey. 1995. "Assessment of Mental Workload with Task-Irrelevant Auditory Probes." *Biol Psychol* 40(1–2):83–100.

Kroehnert, G. 2000. *Basic Training for Trainers: A Handbook for New Trainers*. 3d ed. Sydney: McGraw-Hill.

Kryter, K. D. 1985. *The Effects of Noise on Man*. 2d ed. New York: Academic Press.

Kurke, M. 1961. "Operational Sequence Diagrams in System Design." *Human Factors* 3:66–73.

Lamberts, K., and R. L. Goldstone. 2005. *Handbook of Cognition*. Thousand Oaks, CA: SAGE Publications.

Leigh, D., and D. Leigh. 2006. *The Group Trainer's Handbook: Designing and Delivering Training for Groups*. 3d ed. London: Kogan Page.

Ljunggren, G., S. Dornic, O. Bar-Or, and G. Borg. 1989. *Psychophysics in Action*. Berlin: Springer-Verlag.

Macmillan, N. A., and C. D. Creelman. 2005. *Detection Theory: A User's Guide*, 2d ed. Mahwah, NJ: Lawrence Erlbaum Associates.

Manning, S. A., and E. H. Rosenstock. 1979. *Classical Psychophysics and Scaling*. Huntington, NY: Krieger.

McAfee, R. B., and A. R. Winn. 1989. "The Use of Incentives/Feedback to Enhance Workplace Safety: A Critique of the Literature." *Journal of Safety Research* 20:7–19.

McNicol, D. 1972. *A Primer of Signal Detection Theory*. London: Allen and Unwin.

Miller, G. A. 1956. "The Magical Number Seven Plus or Minus Two: Some Limits on Our Capacity for Processing Information." *Psychol Rev* 63(2):81–97.

Mowbray, H. M., and J. W. Gebhard. 1958. *Man's Senses as Informational Channels* (No. CM-936). Silver Spring, MD: The Johns Hopkins University, Applied Physics Laboratory.

National Aeronautics and Space Administration (NASA). 1995. NASA-STD-3000, *Man- Systems Integration Standards*. Houston: Lyndon B. Johnson Space Center.

Niebel, B. W., and A. Freivalds. 2002. *Methods, Standards and Work Design*. New York: McGraw-Hill.

Norman, D. A. 1988. *The Psychology of Everyday Things*. New York: Basic Books.

Occupational Health and Safety Administration (OSHA). 1974. 29 CFR Part 1910, "Safety and Health Regulations for General Industry." Washington, D.C.: OSHA.

_____. 1974. 29 CFR Part 1910.1030, "Bloodborne Pathogens." Washington, D.C.: OSHA.

O'Donnell, R. D., and F. T. Eggemeier. 1986. "Workload Assessment Methodology," in K. R. Boff, C. Kauffmann, and J. Thomas, eds., *Handbook of Perception and Human Performance*, Cognitive Processes and Performance vol 3. New York: Wiley & Sons, 42/41–42/49.

Peters, R. H. 1989. "Review of Recent Research on Organizational and Behavioral Factors Associated with Mine Safety." Washington, D.C.: U.S. Dept. of the Interior, Bureau of Mines.

Pheasant, S. 1988. *Bodyspace*. London: Taylor & Francis.

Poor, H. V. 1994. *An Introduction to Signal Detection and Estimation*. 2d ed. New York: Springer-Verlag.

Price, H. E., and B. J. Tabachnick. 1968. NASA CR-878, "A Descriptive Model for Determining Optimal Human Performance in Systems. vol 3. An approach for determining the optimal role of man and allocation of functions in an aerospace system (including technical data appendices)." Washington, D.C.: NASA.

Price, H. H. 1985. "The Allocation of Functions in Systems." *Human Factors* 27:33–45.

Psychonomic Society. 1966. *Perception & Psychophysics*. Austin: Psychonomic Society.

Roberts, K., J. F. Burton, M. M. Bodah, and T. Thomason. 2005. *Workplace Injuries and Diseases: Prevention and Compensation: Essays in Honor of Terry Thomason*. Kalamazoo: W.E. Upjohn Institute for Employment Research.

Safir, A., V. Smith, J. Pokorny, and J. L. Brown. 1976. "Optics and Vision Physiology." *Arch Ophthalmol* 94(5):852–862.

Salvendy, G. 1987. *Handbook of Human Factors*. New York: John Wiley.

Sanders, M. S., and E. J. McCormick. 1993. *Human Factors in Engineering and Design*. 7th ed. New York: McGraw-Hill.

Schonhoff, T. A., and A. A. Giordano. 2006. *Detection and Estimation Theory and Its Applications*. Upper Saddle River, NJ: Pearson Prentice Hall.

School Library Manpower Project, and the National Education Association of the United States (NEA). 1969. "Task Analysis Survey Instrument: Definitions of Terms, Checklist of Duties, Status Profile Sheet. A survey instrument of the School Library Manpower Project developed with the Research Division of the National Education Association. Phase I." Chicago: American Association of School Librarians.

Schulz, T., and E. Lessing. 1976. *Hick's Law in a Random Group Design: New Data for Cognitive Interpretations*. Bonn, Germany: Psychologisches Institut der Universität Bonn, Abt. Methodik.

Shanks, D. R. 1997. *Human Memory: A Reader*. New York: St. Martin's Press.

Shepherd, A. 2001. *Hierarchical Task Analysis*. New York: Taylor & Francis.

Stevens, S. S. 1951. *Handbook of Experimental Psychology*. New York: Wiley.

_____. 1957. "On the Psychophysical Law." *Psychological Review* 64(3):153–181.

_____. 1975. *Psychophysics: Introduction to Its Perceptual, Neural, and Social Prospects*. New York: Wiley.

Sum, A. 1999. NCES 1999-470, *Literacy in the Labor Force*. Washington, D.C.: Government Printing Office.

Swets, J. A. 1996. *Signal Detection Theory and ROC Analysis in Psychology and Diagnostics: Collected Papers*. Mahwah, NJ: Lawrence Erlbaum Associates.

Tufts College, Institute for Applied Experimental Psychology, and the United States Office of Naval Research. 1951. *Handbook of Human Engineering Data for Design Engineers (Report)*. Medford, MA: Tufts College.

U.S. Air Force. *Air Force Manual*. Washington, D.C.: Government Printing Office.

U.S. Navy. MIL-STD-1472D, *Human Engineering Design Criteria for Military Systems, Equipment and Facilities*.

Philadelphia: Navy Publishing and Printing Service Office.

Van Cott, H. P., and M. J. Warrick. 1972. No. IP-39, *Man as a System Component*. Washington, D.C.: Government Printing Office.

Verdier, P. A. 1960. *Basic Human Factors for Engineers; The Task Analysis Approach to the Human Engineering of Men and Machines*. New York: Exposition Press.

Waters, T. R., V. Putz-Anderson, A. Garg, and L. J. Fine. 1993. "Revised NIOSH equation for the design and evaluation of manual lifting tasks." *Ergonomics* 36(7):749–776.

Webster, J. C. 1974. "Speech Interference by Noise." Paper presented at the Inter-Noise 74.

Weimer, J. 1993. *Handbook of Ergonomic and Human Factors Tables*. Englewood Cliffs, NJ: Prentice Hall.

Welford, A. T. 1968. *Fundamentals of Skill*. London: Methuen.

Welford, A. T., and J. M. T. Brebner. 1980. *Reaction Times*. New York: Academic Press.

Wickens, C. D. 1984. "Processing Resources in Attention," in *Varieties of Attention* (pp. 63–102). London: Academic Press.

Wickens, C. D., and J. G. Hollands. 2000. *Engineering Psychology and Human Performance*, 3d ed. Upper Saddle River, NJ: Prentice Hall.

Wickens, T. D. 2002. *Elementary Signal Detection Theory*. New York: Oxford University Press.

Wiener, E. L. 1964. "The Performance of Multi-Man Monitoring Teams." *Human Factors* 6:179–184.

_____. 1987. "Application of Vigilance Research: Rare, Medium, or Well Done?" *Human Factors* 29(6):725–736.

Wiener, E. L., R. E. Curry, and M. L. Faustina. 1984. "Vigilance and Task Load: In Search of the Inverted U." *Human Factors* 26(2):215–222.

Wiener, E. L., G. K. Poock, and M. Steele. 1964. "Effect of Time-Sharing on Monitoring Performance: Simple Mental Arithmetic as a Loading Task." *Percept Motor Skills* 19:435–440.

Wierwille, W. W., and F. T. Eggemeier. 1993. "Recommendations for Mental Workload Measurement in a Test and Evaluation Environment." *Human Factors* 35(2):263–281.

Wiker, S. F., G. D. Langolf, and D. B. Chaffin. 1989. "Arm Posture and Human Movement Capability." *Human Factors* 31(4):421–441.

Woodson, W. E. 1981. *Human Factors Design Handbook: Information and Guidelines for the Design of Systems, Facilities, Equipment, and Products for Human Use*. New York: McGraw-Hill.

Woodson, W. E., B. Tillman, and P. Tillman. 1992. *Human Factors Design Handbook: Information and Guidelines for the Design of Systems, Facilities, Equipment, and Products for Human Use*. 2d ed. New York: McGraw-Hill.

Yost, W. A., A. N. Popper, and R. R. Fay. 1993. *Human Psychophysics*. New York: Springer-Verlag.

APPENDIX: RECOMMENDED READINGS

Aasman, J., G. Mulder, and L. J. M. Mulder. 1987. "Operator Effort and the Measurement of Heart-Rate Variability." *Human Factors* 29:161–170.

Adams, J. A. 1982. "Issues in Human Reliability." *Human Factors* 24:1–10.

Bailey, R. W. 1982. *Human Performance Engineering: A Guide for System Designers*. Englewood Cliffs, NJ: Prentice-Hall.

Barnes. R. M. 1949. *Motion and Time Study*. 3d ed. New York: John Wiley & Sons.

Boff, K. R., L. Kaufman, and J. P. Thomas, eds. 1986. *Handbook of Perception and Human Performance, Volume I: Sensory Processes and Perception*. New York: John Wiley & Sons.

_____. 1986. *Handbook of Perception and Human Performance, Volume II: Cognitive Processes and Performance*. New York: John Wiley & Sons.

Boff, K. R., and J. E. Lincoln, eds. 1999. *Engineering Data Compendium: Human Perception and Performance*. Wright-Patterson Air Force Base: Harry G. Armstrong Aerospace Medical Research Laboratory.

Blanchard, B. S., and W. J. Fabrycky. 1990. *Systems Engineering and Analysis*. Englewood Cliffs, NJ: Prentice-Hall.

Brenner, M., E. T. Doherthy, and T. Shipp. 1994. "Speech Measures Indicating Workload Demand." *Aviation, Space, and Environmental Medicine* 65:21–26.

Broadbent, D. E. 1958. *Perception and Communication*. London: Pergamon.

Chapanis, A. 1980. "The Error-Provocative Situation: A Central Measurement Problem in Human Factors Engineering," in W. Tarrants, ed., *The Measurement of Safety Performance*. New York: Garland STPM Press, pp. 99–28.

_____. 1959. *Research Techniques in Human Engineering*. Baltimore: The Johns Hopkins University Press.

_____. 1988. "Words, Words, Words Revisited." *International Review of Ergonomics*. 2:1–30.

Cooper, M. D., and R. A. Phillips. 1994. "Validation of a Safety Climate Measure." Proceedings of the British Psychological Society: 1994 Annual Occupational Psychology Conference. Birmingham, Jan 3–5, 1994.

Cooper, M. D., R. A. Phillips, V. J. Sutherland, and P. J. Makin. 1994. "Reducing Accidents Using Goal-Setting and Feedback: A Field Study." *Journal of Occupational & Organisational Psychology* 67: 219–40.

Crossman, E. R. F. W. 1959. "A Theory of the Acquisition of Speed Skill." *Ergonomics* 2:153–166.

Cushman, W. H., and D. J. Rosenberg. 1991. *Human Factors in Product Design*. Amsterdam: Elsevier.

Department of the Air Force. 1988. DI-CMAN-K8008A, *Data Item Description: System Segment Specification*. Washington, D.C.: Department of the Air Force.

Department of the Army. 1987. Army Regulation 602-2, *Manpower and Personnel Integration (MANPRINT) in Material Acquisition Process*. Washington, D.C.: Department of the Army.

Department of Defense (DOD). 1974. MIL-STD-499A, *Military Standard: Engineering Management*. Washington, D.C.: DOD.

_____. 1985a. MIL-STD-490A, *Specification Practices*. Washington, D.C.: DOD.

_____. 1985b. MIL-STD-1521B, *Technical Reviews and Audits for Systems, Equipment, and Computer Software*. Washington, D.C.: DOD.

_____. 1987. DOD-HDBK-763, *Human Engineering Procedures Guide*. Washington, D.C.: DOD.

_____. 1994. DI-HFAC-80746A, *Human Engineering Design Approach Document—Operator*. Washington, D.C.: DOD.

_____. 1994. DI-HFAC-80747A, *Human Engineering Design Approach Document—Maintainer*. Washington, D.C.: DOD.

_____. 1994. DI-HFAC-80740A, *Human Engineering Program Plan*. Washington, D.C.: DOD.

_____. 1994. DI-HFAC-80745A, *Human Engineering System Analysis Report*. Washington, D.C.: DOD.

_____. 1994. DI-HFAC-80743A, *Human Engineering Test Plan*. Washington, D.C.: DOD.

_____. 1994. DI-HFAC-80744A, *Human Engineering Test Report*. Washington, D.C.: DOD.

Geldard, F. A. 1953. *The Human Senses*. New York: John Wiley & Sons.

Green, A. E., ed. 1982. *High Risk Safety Technology*. New York: Wiley.

Grose, V. L. 1988. *Managing Risk: Systematic Loss Prevention for Executives*. Englewood Cliffs, NJ: Prentice-Hall.

Huey, B. M., and C. D. Wickens, eds. 1993. *Workload Transition*. Washington, D.C.: National Academy Press.

Hughes, P. K., and B. L. Cole. 1988. "The Effect of Attentional Demand on Eye Movement Behaviour When Driving," in *Vision in Vehicles II*. Amsterdam: North-Holland, pp. 221–230.

Itoh, Y, Y. Hayashi, I. Tsukui, and S. Saito. 1990. "The Ergonomic Evaluation of Eye Movement and Mental Workload in Aircraft Pilots." *Ergonomics* 33:719–733.

Johnson, A. K., and E. A. Anderson. 1990. "Stress and Arousal," in J. Cacioppo and L. Tassinary, eds., *Principles of Psychophysiology*. Cambridge: Cambridge University Press, pp. 216–252.

Jones, E. R., R. T. Hennessy, and S. Deutsch, eds. 1985. *Human Factors Aspects of Simulation*. Washington, D.C.: National Academy Press.

Jordan, P. W., and G. I. Johnson. 1993. "Exploring Mental Workload Via TLX: The Case of Operating a Car Stereo Whilst Driving," in A. Gale, ed., *Vision in Vehicles IV*. Amsterdam: North-Holland, pp. 255–262.

Jorna, P. G. A. M. 1992. "Spectral Analysis of Heart Rate and Psychological State: A Review of Its Validity as a Workload Index." *Biological Psychology* 34:237–257.

Kantowitz, B. H. 1987. "Mental Workload," in P. Hancock, ed., *Human Factors Psychology*. Amsterdam: North-Holland, pp. 81–121.

_____. 1992. "Selecting Measures for Human Factors Research." *Human Factors* 34:387–398.

Kaufman, J. E., and H. Haynes, eds. 1981. *IES Lighting Handbook: Reference Volume*. New York: Illuminating Engineering Society of North America.

Kirwan, B. 1987. "Human Reliability Analysis of an Offshore Emergency Blowdown System." *Applied Ergonomics* 18:23–33.

Kryter, K. D. 1972. "Speech Communication," in H. Van Cott and R. Kinkade, eds., *Human Engineering Guide to Equipment Design*. Washington, D.C.: Government Printing Office, pp. 161–226.

Lowrance, W. W. 1976. *Of Acceptable Risk*. Los Altos: William Kaufmann.

Martin, D. K., and S. J. Dain. 1988. "Postural Modifications of VDU Operators Wearing Bifocal Spectacles." *Applied Ergonomics* 19:293–300.

Meister, D. 1989. *Conceptual Aspects of Human Factors*. Baltimore: The Johns Hopkins University Press.

_____. 1971. *Human Factors: Theory and Practice*. New York: John Wiley & Sons.

Muckler, F. A., and S. A. Seven. 1992. "Selecting Performance Measures: 'Objective' Versus 'Subjective' Measurement." *Human Factors* 34:441–455.

National Research Council. 1983. *Risk Assessment in the Federal Government: Managing the Process*. Washington, D.C.: National Academy Press.

Norman, D. A., and D. G. Bobrow. 1975. "On Data-Limited and Resource-Limited Processes." *Cognitive Psychology* 7:44–64.

Parker, J. F. Jr., and V. R. West, eds. 1973. *Bioastronautics Data Book*. Washington, D.C.: Scientific and Technical Information Office, NASA.

Pelsma, K. H., ed. 1987. *Ergonomics Sourcebook: A Guide to Human Factors Information*. Lawrence, KS: The Report Store.

Phillips, C. 2000. *Human Factors Engineering*. New York: John Wiley & Sons.

Rouse, W. B., S. L. Edwards, and J. M. Hammer. 1993. "Modeling the Dynamics of Mental Workload and Human Performance in Complex Systems." *IEEE Transactions on Systems, Man, and Cybernetics* 23:1662–1671.

Rowe, W. D. 1983. *Evaluation Methods for Environmental Standards*. Boca Raton, FL: CRC Press.

Salvendy, G., ed. 2006. *Handbook of Human Factors and Ergonomics*. 3d ed. New York: John Wiley & Sons.

Swain, A. D., and H. K. Guttmann. 1983. Report NUREG/CR-1278, *Handbook of Human Reliability Analysis with Emphasis on Nuclear Power Plant Applications*. Albuquerque: Sandia National Laboratories.

Teigen, K. H. 1994. "Yerkes-Dodson: A Law for All Seasons." *Theory & Psychology* 4:525–547.

Thomson, J. R. 1987. *Engineering Safety Assessment: An Introduction*. New York: Wiley.

Thorsvall, L., and T. Åkerstedt. 1987. "Sleepiness on the Job: Continuously Measured EEG Changes in Train Drivers." *Electroencephalography and Clinical Neurophysiology* 66:502–511.

Tversky, A., and D. Kahneman. 1974. "Judgment Under Uncertainty: Heuristics and Biases." *Science* 185: 1124–1131.

U.S. Army Missile Command. 1984. MIL-H-46855B, *Human Engineering Requirements for Military Systems, Equipment and Facilities*. Redstone Arsenal: U.S. Army Missile Command.

_____. 1989. DOD-HDBK-761A, *Human Engineering Guidelines for Management Information Systems*. Redstone Arsenal: U.S. Army Missile Command.

Van Cott, H. P., and R. G. Kinkade, eds. 1972. *Human Engineering Guide to Equipment Design*, revised ed. Washington, D.C.: Government Printing Office.

Vivoli, G., M. Bergomi, S. Rovesti, G. Carrozzi, and A. Vezzosi. 1993. "Biochemical and Haemodynamic Indicators of Stress in Truck Drivers." *Ergonomics* 36:1089–1097.

5

Benchmarking and Performance Criteria

William Coffey

LEARNING OBJECTIVES

- Outline the necessary elements of a system of metrics.

- Be able to explain the interaction between the formation of metrics and benchmarking.

- Recognize the advantages and disadvantages of leading and lagging metrics, and provide guidelines for the formation of each.

- List the elements of an ergonomics program to be considered for metric formation.

- Be able to explain what a monitoring element is in metric formation and describe how to use such a function in metrics.

- Outline an ergonomic system of metrics based on activities being conducted in the learner's own facility.

THIS CHAPTER was written to provide a basis for consideration as a safety and health professional contemplates the many aspects of measuring performance in the field of ergonomics. Although it does not provide specific instructions, it does outline many of the issues that must be considered when developing a performance measurement system. Many of the metrics that can be considered when measuring the effectiveness of an ergonomics program or general safety program will be highlighted, and many complications that are associated with measuring the performance of systems that may not be purely quantitative will be presented. Although instruction is not provided on how to benchmark against other programs or other companies, some of the essentials for setting up an internal benchmarking effort are provided. Readers will be able to use the information in this chapter as a foundation for their exploration of additional resources that can be used in developing and using sound metrics to measure internal ergonomics program performance, and then benchmarking that performance against outside entities.

Before discussing how to benchmark and measure ergonomics programs, a moment should be taken to discuss benchmarking and metric formation generally. This discussion is of itself involved, but it is worthwhile, because the issues surrounding benchmarking ergonomics mirror the issue of benchmarking in general.

To benchmark is to provide "a point of reference from which measurements may be made" (Webster's 1983). After all, it is hard to change a situation without knowing what it is. Benchmarking is intended to establish a system that can track some feature (in the case of this chapter, safety in general—and ergonomics in particular). Benchmarking is a method for tracking data over time (perhaps at daily, weekly, monthly, or annual intervals).

So why benchmark? Benchmarking gauges (or substantiates) improvements (or deteriorations) in a system. The next questions, of course, are "How do I benchmark?" and "What do I benchmark?" These two questions are two sides of the same issue: developing a system of benchmarking. To continue further in this discussion, another term needs to be introduced, metric. According to *Webster's Dictionary* (1983), a metric is "a standard of measurement." (In this chapter, *benchmarking* indicates a system, and items being benchmarked are called *metrics*.) Certain foundational decisions must be made when developing a benchmarking system. *What metrics will be used?* A metric is not worthwhile unless it provides useful information about the system. Measuring monthly forklift cost fluctuations in an attempt to save money in a 1000–person operation that uses only one forklift will not be effective. *What will be used to benchmark against?* Benchmarking is often thought of as comparing data collected in different times and places about a particular metric (HSE 1999). This means that a unit (perhaps a department, or a facility, or an organization) can compare past data to present data. An excellent example is available by which to highlight the issues above and focus benchmarking and the formation of metrics on safety: the OSHA recordable.

Working Example: OSHA's Record-Keeping Standard

In 1971, the OSHA record-keeping standard, 29 CFR 1904, went into effect. This record-keeping standard applies to all industries that are under the Occupational Safety and Health Administration's (OSHA) purview and requires industrial businesses to track workplace injuries (OSHA 1971). It explains which injuries are to be recorded and categorizes injuries into types (such as recordable, restricted-duty, fatalities, and those causing lost work days). Injuries are to be tracked annually and the totals posted after the end of each year. The OSHA record-keeping standard created the largest U.S.–based benchmarking system ever used and made the OSHA recordable rate the most widely used metric in the United States (and made the OSHA record-keeping standard the most comprehensive benchmarking system)—simply because the majority of industrial businesses, as well as civil employers, must comply with OSHA's record-keeping standard.[1] The system addresses all the issues raised above.

In theory, benchmarking is done to reduce employee injuries. In the case of the record-keeping standard, benchmarking defines the types of injuries that are *recordable* and tallies them yearly. Benchmarking is performed by all industries regulated by OSHA. In fact, the Bureau of Labor Statistics (BLS) tracks recordable rates by Standard Industry Codes (SIC) or by the newer North American Industry Classification System (NAICS) (United States, 2002). The data can be broken down to look at injuries by facility and company size, by general occupational type, by parts of body affected, or by types of event or exposure (BLS 2006). This gives companies an opportunity to benchmark against other companies in the NAICS, against similarly sized companies, against their industry in general, or just against their own data from previous years. By using a normalizing formula (the incident rate calculation), facilities of varying size and function can (theoretically) be compared (OSHA 2005). This data is scrutinized by governmental agencies, corporate offices, union committees, and all other parties who are interested in safety. Facilities even go so far as to track these metrics monthly, posting graphs or other visual aids to represent occurrences. OSHA has begun to use this data to decide which sites to inspect (OSHA 2006a).

Benchmarking and the formation of metrics are only tools—not ends in themselves. A system's quality is reflected in its success in meeting its objectives—in the case of safety, lowering injury rates. Benchmarking, as any tool, is not perfect. A benchmarking system that is not developed correctly will not meet goals. Poorly correlating comparison groups (such as a manufacturing site and an administrative office), creating too many metrics to track effectively, and developing inappropriate metrics can all detract from a benchmarking system's usefulness. OSHA's record-keeping system illustrates this point.

The appropriateness of 29 CFR 1904 has been debated. Although OSHA made efforts to improve the system, leading to the 2000 revision of the standard (OSHA 1971), detractors of the system point to its persisting deficiencies (Nash 2001). OSHA's recordability rules must be applied consistently in order to be effective, but interpretations of what is, and what is not, recordable are still debated. Some subjectivity exists as regards attendant physicians, as well as some variability depending on the nature of specific industries. An injury that does not affect one employee's ability to work at one employment may make another employee in another job unable to work (OSHA 2006b).

This chapter is neither an affirmation of nor an attack upon the OSHA record-keeping system. Rather, it is intended to present the general concepts of benchmarking and metric formulation. A comprehensive analysis of the system is well beyond the scope of this chapter and inconsistent with its purpose.

Metric Classification

Metrics, the gauges of systems, can be divided into two main types: pre-indication (leading) and post-indication (lagging). Pre-indication metrics are qualities that can be measured prior to an occurrence. Post-indication metrics are qualities that are measured after an occurrence—including the occurrence itself. The OSHA recordable illustrates this concept, being itself a post-indication metric. When an injury occurs, it is recorded on an OSHA 300 log according to the criteria it meets. Comparing amounts of workers' compensation dollars spent yearly is also a post-indication metric—something measured after it has occurred. In other words, to monitor injuries, an employer measures the number that occur, later remeasuring in the hope of finding that some reduction has occurred.

Because of the diversity of industry safety programs, however, pre-indication metrics are not nearly so clear-cut. They can, for example, track the completion of audit corrective actions (to ensure that unsafe conditions are corrected before injuries occur) or track the percentage of initial training that is conducted (to ensure that new employees are trained how to work safely before they are injured). Pre-indication metrics are used by analysts to discover which factors will lower the probabilities of injuries' occurring, ensuring that these factors are then controlled.

Decisions about metrics—specifically about pre- and post-indication metrics—bring evaluators face to face with the confounding factors of time, resources, and complexity. Post-indication metrics are direct, observable results that are either outright goals or that correlate very closely with goals. Because no organization has infinite resources, much of the creation of post-indication metrics boils down to resource allocation. A new training program may show reductions in recordable rates and workers' compensation dollars for the year *after* it is implemented—but how can an evaluator know whether resources being applied *now* will affect future rates? No one wants to continue to expend resources in ways that will not help in meeting goals.

This is where developing pre-indication metrics may be helpful. Post-indication metrics are observable (they can be seen and measured in reference to goals), but pre-indication metrics are predictive (their measurements help predict whether certain goals will be met in the future). Pre-indication metrics take measurements while resources are being expended in order to predict the effectiveness of such expenditures. The chief difficulty associated with pre-indication metrics is deciding what to measure. Evaluators must decide which program elements will likely increase safety before implementing them. A perfect world would see the implementation of a safety-system "improvement," the tracking and measuring of some observable element of the program, and a later evaluation of whether the "improvement" decreased the occurrence of injuries. If so, the program should be continued. If not, the process should be repeated with a new safety-system "improvement." But the real world is not so forgiving. In reality, companies develop lists of possible improvements, prioritize them, and implement as well as possible those improvements most likely to decrease injury rates (a practice referred to as risk assessment). The amount and types of improvements implemented are governed by the resources available. One simple improvement can be implemented or multiple complex improvements can be implemented. An attribute of the program can be selected and tracked to ensure that the program is being implemented and maintained.

difficulties when working with ...cs is choosing measurable quan... predictive. A company that has an ...rogram may find by monitoring and ...int of audits conducted that the more a... ...as, the lower its injury rates. The company woul... ...n set metrics for auditing: *the more audits the facility conducts, the fewer its injuries*. But the difficulty of pre-indication metrics is that they can be organizationally specific. Auditing and tracking audits is a worthwhile safety activity, but in some organizations it may not have the same significance as in others. Take, for example, Company B, which has noticed the association of injury reductions with the auditing metrics of Company A, described above. Company B decides to use the same metrics, even though it does not have the extensive auditing system of Company A. An auditing metric may not be the best predictor of goal attainment for Company B, which has no reason to expect a correlation between auditing and injury reduction.

Causality can be another confounding issue. Company B assumes that the auditing program itself decreases injury rates; but perhaps Company A's program includes aggressive corrective action, incorporating strict criteria for when corrective actions must be completed. This could have a great effect on the reduction of injuries—but these factors may not have been taken into account by Company B when it tracks the number of audits it conducts. Tracking audits may not have anything to do with injury reduction in Company B.

Another difficulty associated with pre-indication metrics is the impossibility of proving a null hypothesis (Streiner 2003). In other words, "How can a negative be proven?" Relating pre- to post-indication metrics can be difficult. How can anyone quantify how many injuries *didn't* happen? Even when post-indication metrics are not within goals (as when recordable rates are high), how can it be truly ascertained that rates would have been higher if not for the efforts measured by the metrics?

MUSCULOSKELETAL DISORDERS (MSDs)

The discussion on benchmarking and metric formation in general extends to ergonomics in the specific. In fact, some of the issues discussed are compounded. Ergonomic injuries (referred to as musculoskeletal disorders, or MSDs) do not happen immediately but develop over time. In fact, some can take years to develop (NIOSH 1997). How can post-indication metrics (that is, injury rates) alone indicate whether improvements to the safety system are effective? The injuries recorded have developed over years and are only recognized when their symptoms interfere with everyday activities. Even if drastic efforts are taken after the diagnosis of the first few occurrences, further injuries will be diagnosed and recorded; such injuries develop over many years and the first have only now been recognized. On the other hand, poor pre-indication metrics may cause resources to be put into a safety program for years despite a complete lack of indication of the effectiveness of doing so. In ergonomics both post- and pre-indication metrics should be developed. Post-indication metrics are direct measurements that can include number of MSD recordables, number of repetitive motion injuries, and total of worker compensation dollars spent because of MSDs. But what are used as pre-indication metrics? Such items must have predictive value—and in the field of ergonomics, this is common. Risk factors, pain surveys, workplace design, and training are all possible fixes for ergonomic issues and can be used as pre-indication metrics. Suppose a company is implementing a new ergonomics program. The safety manager has discovered that the company is currently suffering five MSD injuries per 100 workers yearly. The company, which is paying about $175,000 annually because of such injuries, wants to put in place an ergonomics program to reduce these numbers and has set a goal of a 40 percent reduction within three years. The number of MSDs and the workers' compensation dollars are post-indication metrics and are used as a baseline against which to compare. As part of the ergonomics improvement program, the following elements are implemented: ergonomic task evaluation, new process ergonomic design review, worker training in lifting and repetitive motion, and the formation of an ergonomics committee. Pre-indication metrics are based on these elements. The metrics for the first one-year period may require that (1) all facilities will have formed an ergonomics committee, (2) all facilities will have conducted general training in ergonomics for their associates, (3) a new ergonomic process-review procedure will have been developed and implemented, and (4) a minimum of five work-task analyses will have been completed. These example metrics will depend greatly upon the organization's culture and the resources available. But even these examples emphasize a key factor: shifting metrics. These metrics may not have meaning in the future—something that can be true of both pre- and post-indication metrics. Recordable injuries may be reduced to nonexistence. Tracking recordables at this point may not be necessary. Measuring the amount of first-aid cases may be a more useful alternative—see the Hendricks pyramid (Heinrich 1931).

METRIC FORMATION

When developing metrics, the following stipulations should be noted:

1. Pre-indication metrics, post-indication metrics, and goals must be decided upon and the differences between them understood. A goal is a desired outcome for a metric in a certain timeframe—such as a 10 percent reduction in recordable injuries during a 2-year period.
2. Metrics should be quantifiable and measurable.
3. Pre-indication metrics should address interventions used to improve post-indication metrics.
4. As goals are met, new metrics and goals should be introduced that reflect the current situation.

Injury/Complaint Trending

Part of the intent of the OSHA record-keeping standard was to reduce the frequency of worker injuries. But how is writing down when people are hurt going to do this? The idea was that when injuries are recorded, problem areas and jobs will be evident (OSHA 1971). These areas and jobs are where the most attention to safety must be paid. This is trend analysis—the use of data about injuries to highlight priority areas where safety attention is needed. A set of data is analyzed to discover whether reoccurrences are so numerous that they constitute a pattern. Although OSHA's record-keeping standard is a good example, trend analysis can extend to any data set.

There are often unrecognized patterns in any system. Analysts try to find these patterns by means of trend analysis. But what is the point of finding a pattern of injuries? Two advantages are clearly evident: credibility and priority. Both of these advantages stem from predictabilities. A properly conducted trend analysis enables the prediction of yet-to-come occurrences (Wikipedia 2007a).

By understanding injuries, safety professionals can prioritize their safety efforts. Because they are often spread thin in the expenditure of their resources, money and time used on one project must often be removed to another project. From an economic standpoint, the cost of a project is not only the time and money used to complete it but also the loss of benefits from other projects that could have used that time and money. Accurate trend analysis allows the areas having the most severe injury issues to be identified, allowing for the development of programs to abate the associated hazards—but safety professionals must be cautious. The prioritization of trend-analysis data must occur case by case, and broad statements should be kept at a minimum. The frequency, severity, and possibility of fatality from injuries are all possible ways to prioritize resource expenditure. Arguments can be made for or against prioritizing any one of the above. The greater the frequency of injuries, the more people are getting hurt, and the higher is the probability of a fatal injury. But what about numerous minor injuries, accompanied by a smaller percentage of very severe injuries? Shouldn't these severe injuries be addressed? All safety professionals (in fact, all business professionals) hold the prevention of fatalities as a high priority, but what happens when the data do not lend themselves to the prediction of fatalities? Most of the time, evident trends will themselves set the priorities. Consider the OSHA 300 log of a small, 50-employee manufacturer. The last three years of the OSHA log shows 18 recordable injuries—12 cuts to hands, 1 back injury, 1 case of heat exhaustion, 1 instance of hearing loss, 1 foot injury, 1 case of skin irritation, and 1 case of carpal tunnel syndrome. Hand and back injuries might be made priorities by some safety professionals: hand injuries because they constitute the bulk of this company's recordables for the last three years and back injuries because of their severity, being some of the most disabling and costly injuries. This data was analyzed by type of body part injured. Any of the data types on the OSHA 300 form (organized by employee, by date, by department, by job title, by body part, or by type) can be analyzed (OSHA 1971). Sorting the data by each category is a good place to begin a trend analysis. In the trend analysis above, the number of injuries (first criteria) is being sorted versus the body part affected (second criteria). Larger data sets may require the further expansion of these criteria. The first criteria could be the number of recordable injuries and the second could be organization by body part, and then the data could be re-sorted in these subsets by department. The first requirement of metric evaluation is the existence of a base of data broad enough to provide significant results.

Credibility is not an issue that is well defined or frequently used in trend analysis. The connection it has with prioritizing, however, is worth mentioning. To develop and implement programs to correct the hazards and decrease injuries requires resources (mostly of time and money). Convincing others of the effectiveness of expending these resources can be a daunting task. Trend analysis provides a foundation for justifying resource allocation. In fact, the trend analysis itself sets the stage for benchmarking and the formation of metrics. Trend analysis is simply self-benchmarking. When trend analysis is accurate and future outcomes can begin to be predicted, an organization can gain the ability to change outcomes by expending resources in an informed manner. If a trend can show how, after resources are applied, a predicted change occurred, the credibility of safety professionals' recommendations for the expenditure of resources increases, making it easier to get approval for the allocation of resources.

As with other systems, the devil is in the details. Trend analyses can be informal, comparing two or three factors and representing them visually (charting or graphing), or formal (involving t-square analysis or normal distributions). The intended use of data, the mathematical skill of analysts, and the nature and extent of the data available will all affect the formality of trend analysis. Formal trend analysis that depends upon statistical principles will be affected by the interaction of sample sizes with the desired types of analysis and the desired level of uncertainty (Wikipedia 2007b). In fact, sample size plays a part even in informal trend analysis. When a frequency of occurrence is cited, and the sample size is small or the time period is short, it will be difficult—if not impossible—to draw predictable conclusions because of the marked increase in the effects of chance and variation (Wikipedia 2007b). Probabilities simply may not have had enough opportunity to manifest themselves. When a die is rolled six times, a high probability exists of it rolling a 1 at least once. But if a die is rolled only twice, that probability is much lower and (though present) may not manifest itself. But what happens when a probability is not known—when the equivalent of a four-, six-, ten-, twelve-, or twenty-sided die is rolled? Each of these approaches closer to the use of trend analysis to make predictions. If in twenty rolls five 1s are rolled, it is likely that a four-sided die is being rolled. If in eighteen rolls three 1s are rolled, a six-sided die seems likely—and so on. As in the case of trend analysis, observations (the numbers and results of die rolls) are used to evaluate probabilities. Such predictions make possible increasingly informed decisions about the efficient expenditure of resources. Remember, however, that as sample sizes increase, data should be more standardized. If only 20 recordable cases are compared, a fair amount of detail can be attached to each case; but if 2000 cases are compared, the information should be standardized to make comparison efficient and effective.

These methods are well-suited to the field of ergonomics. Because OSHA 300 injury logs, as the most common form of injury tracking in the United States, are also the most common tool for trend analysis, a definition of what injuries are ergonomic concerns is a good starting point. Ergonomics is the science of fitting each work area to the employees using it (thereby reducing stresses on them) (NSC 1997a). "Ergonomic injury," of course, is a misnomer—ergonomics is the cure, not the problem. Various names exist for the injuries that ergonomics can abate. Cumulative trauma disorders (CTDs) (NSC 1977b), repetitive strain injuries (RSIs) (NHS 2006), and musculoskeletal disorders (MSDs) (OR-OSHA 2005, 8) are all terms used to refer to disorders that affect muscles, tendons, ligaments, nerves, and blood vessels. General examples of these include strains, sprains, pulled muscles, back injuries. Specific conditions include tenosynovitis, tendonitis, epicondylitis, and carpal tunnel syndrome. For simplicity's sake, the acronym MSD will be used for the remainder of this chapter to refer to *ergonomic injuries*. OSHA included a MSD column on the OSHA 300 form in its original update to 29 CFR 1904 but subsequently removed it (OSHA 2001). The performance of this trend analysis will depend on the OSHA 300 log used; the number of occurrences will guide the trend analysis. The confounding factor here is the same for ergonomics as for performance measurement in general: the long onset time of injury. If, as Heinrich says (1931), minor injuries and near-misses lead up to major injuries (those considered *recordable*), will this not also be true for cases of ergonomic injuries? If these factors can be tracked, logic and Heinrich seem to dictate that they should be used to indicate the success or failure of efforts at safety. By tracking precursors, safety professionals can adjust or change programs as needed. But what precursors can be used? Two such items seem particularly applicable: complaints and risk factors.

Are there true *near-misses* for ergonomic injuries? Their nature would make them seem unlikely to point to an occurrence as "almost an injury." But ergonomics does have a near-miss analogy: complaints of pain and discomfort. In a chronic injury, an area is repeatedly injured beyond its ability to effectively repair itself. Infrequent occurrences are healed by the body, but consistent reinjury leads to chronic disease, and to pain or discomfort in the employee, including complaints of burning sensations, cramping, stiffness, and pain, among other symptoms. These complaints can be helpful indicators in benchmarking ergonomics. Unfortunately, these symptoms are often overlooked or ignored as *normal*. Just as not all employers track near-misses, so there may not be a system in place to track these complaints—an excellent opportunity entirely missed. One example of complaint tracking can be instituting a complaint form in an area with known ergonomic hazards. By counting the number and types of complaints after the implementation of various improvements, indications are given of the success or failure of these initiatives. The amount of complaints per department can be compared to provide some indication of where to set priorities. Note that personality types may strongly affect levels of complaints. One subset of the population may be reticent to complain (thinking, *I'm just getting old—this isn't anything big*), and another may be more vocal with its complaints, overshadowing those of others.

Risk (or contributing) factors are another area open to trend analysis (OR-OSHA 2005, 18). Risk factors are those conditions that may contribute to ergonomic injuries. The higher the number of risk factors associated with a particular task, the more likely is the occurrence of an ergonomic injury. By tracking the number of risk factors per job (or task, or department), the efforts of reduction programs can be evaluated and, if necessary, prioritized as the result of comparisons. Anything that is shown to be a precursor to an ergonomic injury and is able to be tracked can be used for trend analysis.

Workers' Compensation Loss

When benchmarking, one of the more useful things to consider can be workers' compensation data. The ethical and moral reasons for effective ergonomics programs are many, and they are supplemented by significant economic reasons. Workers' compensation data indicate the actual cost (to companies or insurance carriers) per injury and provide a way to gauge how injuries are directly (economically) affecting a company. Although workers' compensation data seem suitable for benchmarking, some cautions should be noted. The most basic workers' compensation data are a reflection of cost (money spent). Although workers' compensation data can be useful to a general safety program, the definition of *ergonomic injury* should be clear. Data must be compared around a certain subset of injuries, as opposed to the whole.

How a company's workers' compensation program is administered can also be a factor. Workers' compensation data can be affected more by the case-management skills of a facility than by the occurrence and severity of injuries. A company's ability to offer transitional work (light duty) can also play a part. Two companies can have identical incident and severity rates for ergonomic injuries, but one may have a very aggressive transitional work program, making the workers' compensation dollars spent by the company with the transitional work program lower because of reduction in lost-time indemnity.

Even workers' compensation data itself may be misleading because the data do not reflect the indirect costs of incidents. Arguments have been made that the indirect costs of injuries (lost efficiency, employee morale, training time) can be 4 to 12 times greater than the direct costs, such as medical bills and indemnity (OR-OSHA 2005).

Despite these cautions, workers' compensation data can provide an array of benchmarking opportunities. Workers' compensation data move away from reliance on the OSHA recordable and are collected about any injury that generates expenses. Although first aid and first-aid supplies used on site are not factored into these data, including these costs is not difficult.

These data are also a bottom-line comparison and can offer an opportunity to show the company real cost savings from a successful program. These cost savings can take the form of lower premiums for a company insurance policy. Companies that insure themselves for Workers' Compensation may see even greater savings. Not only is the company saving on direct and indirect costs, it is also avoiding tying up its money in reserve (money set aside as the expected costs of workers' compensation cases). The workers' compensation administrator or the risk-management department can be asked to provide injury data broken down by types of injury, body parts injured, dates and times of injuries, work days lost because of injury, and more. These data are collected for any injury incurring financial expenses and can greatly aid metric formation and trend analysis.

Job/Task Analysis

Some good lagging metrics have been discussed, but leading metrics need the same sort of examination. Task analysis is a tool used in ergonomic assessments and improvements (NIOSH 1997). For the purpose of this chapter, an analyst performing a task analysis normally looks at a *job* (a job title, a specific department, or an individual responsibility) in its totality before breaking it down into specific tasks to be performed. These tasks are then analyzed for MSD hazards, and ergonomic corrective actions are advised. Consider a machine operator who has five responsibilities, or tasks, making up his or her job:

1. Ensuring that the machine continues running
2. Clearing part jams
3. Quality-checking a certain amount of parts
4. Loading raw material
5. Housekeeping of the general area

An ergonomic analysis should identify MSD hazards and aid in developing work techniques—perhaps even indicate where to install equipment—for minimizing or eliminating any hazards (which are often prioritized).

If task analysis is a corrective measure, how can it be a metric? To answer this question, leading metrics must be discussed. Leading metrics, though not absolute, are predictive. Because task analysis is an ergonomics' tool, the corrective actions that follow it can be used as metrics (Sanders and McCormick 1993).

In its simplest form, a corporation could structure an ergonomic metric as follows:

All facilities in the corporation will:

1. Identify 100 percent of the work tasks performed in the facility and develop task analysis for 25 percent of the tasks in the first year.
2. Complete 75 percent of the task analysis for the facility in the second year.
3. Develop 100 percent of the task analysis by the third year.
4. Complete all corrective actions within 60 days of being generated.

If this were the company's only ergonomic metric, it would be lacking. Ergonomics, like safety in general, does not have a *one-size-fits-all* answer. Task analysis is a tool for ergonomic improvement, but it is not itself an ergonomics program. Other ergonomic elements would also need to have metrics developed. If, however, a company were developing a metric system for safety in general and wanted a single metric as a barometer for ergonomics, this metric might be acceptable. Again, it depends on the goal. It is possible to have too many metrics to follow; what is to be measured must be decided. If a combination of leading and lagging metrics is desired for indication of the progress of a safety program as a whole, and the entire scheme is such that a single metric on ergonomics is desired, the above may be adequate. If the desire is to develop a metric system specific to the ergonomics program, additional metrics should be developed that will center on the actions taken to improve ergonomics.

For some companies, the metric above may be too rigid. For a corporation made up of small manufacturing sites, it might be appropriate, but a company with facilities employing from 20 to 2000 employees, and having manufacturing and service sectors in global locations, may have trouble making this specific metric work.

The basic principles, however, can be adapted for other cases. In the metric above, an objective and a timeframe are set forth. When establishing a metric, the hazards involved must be identified. Because this is true of any leading metric, it is a good place to start. A fair amount of time should be devoted to this identification process for two reasons. The first has to do with thoroughness. An understanding at the beginning of the project of all the hazards involved will allow for knowledgeable prioritization. Any additional hazards that are discovered will require the original plan to be adapted, decreasing efficiency. The second reason is one of logistics. The facilities that have to comply with the new metrics will most likely need resources to aid them in doing so. By providing time during the identification phase of the metric, the facilities can develop schemes for obtaining the resources they will need, whether consultants, training, or internal corporate expertise—and all of these methods require funding. Consultants charge fees, training classes have costs (for both the training and possibly travel), and charge-back costs may accompany corporate aid. By allowing extra time during the identification phase, facilities will be able to develop budgets in support of ergonomic activities.

One other issue that must be decided during the identification phase is how to define a *task*. There is no clear-cut answer to this question. The *job* must be evaluated with an eye to recognizing its basic elements as distinct units. Think about the task list of the machine operator given a moment ago. Are the tasks listed adequate for the purpose of analysis? Perhaps, but further depth would be needed. The following may, in fact, be true:

1. Ensure that the machine continues running: strictly observe the machine.
2. Clear part jams: power down the machine, initiate lockout/tagout, physically (expending moderate effort) remove unfinished parts.
3. Perform quality checks on a certain amount of parts: nonstrenuously remove and measure five parts hourly.
4. Load raw material: raise a mechanical drum hoist.
5. Housekeep the general area: sweep and remove light debris.

But what if the job were changed slightly? What if the operator must retrieve drums of parts (with a forklift) and place them onto the drum hoist? What if housekeeping included machine maintenance? What if maintaining the machine includes lubricating it? All of these items will expand the task list.

A counterpoint to the above consideration is that a hazard identified for one *job* may be applicable to other jobs, limiting the necessity of repeated work for analysts. If all workers in a facility who are required to use a forklift do so in a similar manner, one task analysis for forklift use applies to those employees. Such common tasks can be prioritized for analysis, creating benefits for all employees very efficiently.

Once the identification phase is completed, further metrics can be developed around the task analysis. Amounts to be completed in set times (as in the example above) need not be percentages but may be specific numbers (such as ten task analyses completed in one year). Notice, however, that all the examples have a timeframe. Yearly repetition is convenient but is certainly adjustable to the specific needs and logistic issues of a company. If a company feels that the ability and resources are present to conduct task analyses for *all* jobs simultaneously, the use of percentages would be appropriate.

The task-analysis metric can be taken further. Once a facility has 100 percent of its tasks analyzed, what next? A metric requiring the facility to develop and implement a task-analysis review for all new projects and all modifications to work areas can be used to ensure continuation of ergonomic analysis to all work areas. Most importantly, when general hazards, as well as their causes, have been identified, the metric process can be continued as the prevention process.

Reduction Efforts

The task analysis above is actually an extended identification method that aims to identify ergonomic hazards, after which various reduction efforts come into play. What reduction methods will be used is an issue best decided by the specific approach chosen to address the problem. Whatever methods are chosen, metrics can be based on them.

Task analysis can be tied to reduction efforts, and simple metrics developed. The example used above can be extended even without a comprehensive plan to address ergonomics. A modified example could read as follows:

1. By the conclusion of the first year, have identified 100 percent of the ergonomic hazards, prioritized them, and conducted task analysis for 10 percent of them.
2. By the conclusion of the second year, have conducted task analysis for 35 percent of the hazards and developed corrective actions based on all completed task analysis.
3. By the conclusion of the third year, have completed task analysis for 55 percent of the hazards and developed corrective actions based on all completed task analysis.

This example can be continued. At root, some reduction effort is required to address whatever hazards are found; but it is up to each facility to decide how to correct them.

What if the facility above has decided that work area redesign should be used to eliminate the ergonomic hazard? A simple method, as above, can be used, or other factors can start affecting the metric. What if a metric is desired that takes frequency into account? The company in the above example wishes to use its resources to address ergonomic issues efficiently. The problem with the example above is that it assumes that all hazards can be addressed and that all corrective actions will be effective. Neither of these may be the case. Laying out the metric as above does not recognize that no effective fixes may exist for certain issues. Furthermore, when a corrective action "makes employees aware," is it truly effective? Can one say that a corrective action was implemented when no other course of action seemed appropriate? What if the severity is tied to the metric? Then the following might be effective:

- Using the facility OSHA 300 logs, prioritize the jobs having the highest ergonomic incident rates.
- Develop corrective actions (in this case process redesign) for the three tasks having the highest incident rates. Monitor these task symptoms during the first year.
- During the second year, develop redesigns for the tasks having the next four highest rates of incidence. Monitor these task symptoms, comparing them to the original three tasks from the year before and noting any reduction or increase.
- During the third year, develop redesigns for the tasks having the next four highest rates of incidence. Monitor these task symptoms, comparing their symptom rates to those of the task redesigns completed during previous years and noting any reduction or increase in symptoms.

The metric above is attempting to be efficient (by concentrating on the areas of most frequent injury), progressive (performing additional corrective actions yearly), and relevant (monitoring symptoms to evaluate the accuracy of the leading metric's predictions). Although the above scheme may be appropriate only for a limited number of organizations, its base concepts can provide guidance in forming metrics in general. The metric above tries to focus the company's efforts. No organization has unlimited resources.

In the first example above, the company decides which tasks to prioritize for corrective actions. Nevertheless, influences outside the safety realm may take precedence. Workload can cause a person to implement a less time-intensive and less effective corrective action in order to meet metric timelines. A leading metric can be met in such a way that it has no effect on a lagging metric. So long as multiple pressures exist in the workplace, the formation of leading metrics must take into account the final outcome desired (decreasing the number of injuries).

The second (more complex) example takes this into account, instructing the facilities to quantifiably correct certain hazards first (by using the OSHA 300 logs to determine the most hazardous tasks). If necessary, a facility can be called upon to justify its choice of corrective actions. The other advantage of the second example is its monitoring of symptoms. Leading metrics should be predictive; ergonomic injuries are years in the making. When symptoms are monitored, determining whether corrective actions are improving a situation can be accomplished to some extent. If symptoms associated with MSDs are not decreasing in frequency, the corrective action can be analyzed and, if determined to be ineffective, reworked. This monitoring also serves another purpose. Monitoring can assure that hazards are being addressed, but it can also indicate a systematic problem. It is likely that some corrective actions will have to be reworked, but what about a facility that reworks the majority of its corrective actions? Might this facility not have difficulty in developing effective corrective actions? The metric described above could not indicate whether such problems were related in particular to skills, management support, implementation, or any number of other factors, but it would emphasize the inefficient use of resources, identifying this area for analysis in order to discover the cause. One of the purposes of metrics is to ensure that limited resources are used meaningfully.

Both of the examples above are concrete and quantifiable. But what about something less quantifiable, such as a voluntary stretching program? How can metrics be developed around something like this? It can be assumed that an analysis was conducted and that it was determined that such a program would be a benefit. An oversimplified metric could read as follows:

> A voluntary stretching program will be developed that will have 85 percent employee participation.

This metric is rather vague and somewhat contradictory. How can 85 percent participation be ensured for a voluntary activity? A more quantifiable metric might read as follows:

- A voluntary, preshift stretching program will be developed. Employees will be encouraged to attend the stretching sessions five minutes before their shifts start.
- Symptoms will be monitored throughout the facility and those of employees who stretch will be compared to those of employees who do not.
- If after two years, employee participation is not at 50 percent and no mitigation of the symptoms of employees participating in the stretching program is evident, the program will be reevaluated to decide whether to continue it.

This metric could be used even if other reduction efforts are being used by changing the last criteria to read:

- If after two years, employee participation is not at 50 percent and a greater mitigation of symptom reduction is not noted in employees participating in the stretching program, the program will be reevaluated to decide whether to continue it.

This metric has some advantages. By means of monitoring, the effectiveness of the metric is recognizable, and a criterion is built in to decide on its continuation (this is especially important, because this program requires ongoing resources during its continuance). If the program is ineffective, the resources being expended on its behalf can be reassigned.

Diagnostic Tools

So far, the discussion on metrics has centered on either direct effects (such as injuries or dollars spent) or corrective plans (such as hazard identification or corrective actions). Are there any additional quantifiable ways to measure progress in ergonomics? Yes, when delving into occupational medicine. Because ergonomics centers on MSDs, methods that measure stress on the musculoskeletal system may act as beneficial metrics. Some of the diagnostic tools used in occupational medicine may fill the bill for metrics. However, several issues exist when using these tools as metrics. All of the methods mentioned require some level of training for effective use. Some of the methods are complicated and require input from licensed medical personnel. The time, money, or infringement of privacy required by some methods makes them prohibitive for use in the general employee population. With these things understood, however, applications do exist for these methods as metrics.

Job descriptions can be effective tools when they include not just verbal descriptions of the jobs themselves but also of the physical requirements for performing this job in a satisfactory manner. Such job descriptions include requirements to "lift 75 pounds," "stand for approximately 6 hours per shift," or "repeatedly lift over the head boxes weighing up to 50 pounds." These descriptions are quantifiable, stating the physical exertion required. An occupational medicine professional can help with the formation of a facility's job descriptions; but, with training, anyone can develop job descriptions (OSHA 2002). These job descriptions help return injured employees to work and serve as aids in hazard identification (the job descriptions having the highest weight limits can be flagged as hazardous). A metric can be developed around job descriptions requiring that, say, within two years the facility will have job descriptions for 100 percent of its jobs. Job descriptions can be used even more aggressively. Per NIOSH, employees should not be lifting loads over 51 pounds (NIOSH 1994, 13). A metric can be constructed that reads as follows:

- In the first year, job descriptions will be developed for all jobs in a facility and any job requiring employees to lift more than 51 pounds will be flagged.
- In the second year, the facility will prioritize all jobs requiring the lifting of more than 51 pounds, developing corrective actions for reducing the lifted weight or lift frequency for 20 percent of these jobs.
- In the third year, the facility will have reduced the lifted weight or lift frequency for 50 percent of all jobs requiring employees to lift more than 51 pounds.

This metric is quantifiable, and a monitoring function can be developed to determine whether its corrective actions are effective.

Preemployment screening (such as the Matheson System) is another tool that may serve a metric function (Matheson Discussion Group). Preemployment screenings are an extension of job descriptions. After a job description is developed, an occupational medicine professional can develop a questionnaire of past medical history and a series of functional *tests* (evaluating the employee's ability to reach, stretch, and lift) based on the job requirements. These screenings are fairly basic, and past injuries and illnesses play a large factor in them. Metrics formed from preemployment screenings will be similar to those developed from job descriptions.

Functional capacity evaluation (FCE), an extension of preemployment screening, normally is performed in connection with workers' compensation (Matheson). It measures impairment. Employees are put through a series of lifting, carrying, and moving tests that mimic job tasks. The testing is performed in an array of positions to simulate the work environment and to isolate specific muscles and muscle groups. Limb and back position determine the muscles used to perform work. If an injured muscle is isolated, it may not be able to handle the workload of a task as if it were one of a number used in the task. The FCE then outlines the capacities of the injured employee for use in the return-to-work program (RTW) and in the assignment of impairment ratings (California Department of Industrial Relations). An impairment rating is used to determine how *disabled* an employee is and is often the basis of a cash award for an injured employee. FCEs are normally performed a limited number of times; multiple performance reflects the possibility of some change in the level of impairment. Specific muscle-group testing can be conducted in addition to FCEs, discovering the capabilities of specific muscles or groups of muscles (unlike FCEs, which look at a worker's total abilities).

How can these tests be incorporated into metrics? In the way they are classically conducted, they are not effective; but with some modification, they may be of value. One of the objectives of the formation of metrics is the building of quantifiable metrics. When using a task analysis, this may at times be subjective. The above tests enable one the quantification of information from a task analysis. The hazards found during the hazard analysis can be measured for the work required to be performed. By incorporating the data on force into the task analysis, the tasks can be quantified.

These tests can also serve as monitoring functions. The force required to do a job can be measured initially and then again, after a reduction effort has been implemented. By implementing a reduction measure, some of the force required to perform the task should be reduced. A good example of this can be seen in measuring the force required to use certain hand tools. In one study, the force required to use an *ergonomically* designed pair of pliers was greater than the force required to use an ordinary pair (UC Berkeley ME). The ergonomic value of such a tool is questionable. Furthermore, the *ergonomically* designed tool cost more than an ordinary tool, making the *ergonomic* choice an inefficient use of resources.

Diagnostic tools help in the quantification of metrics, giving the ability to quantify, measure, and prioritize the use of resources.

Administrative Functions

Engineering controls are the preferred methods of hazard control (NSC 1997, 124). As long as engineering controls are in place, hazards are all but eliminated. Sometimes, however, engineering controls are not possible and must be replaced by the second level of the hierarchy—administrative control. Administrative control mitigates hazards by controlling how tasks are done, making hazards more dependent on human behavior. Can effective metrics be developed for behavior-driven items? Yes, and three common administrative ergonomic controls can be given as examples: training in ergonomics (NIOSH 1997, 34), ergonomics committees (NIOSH 1997, 8) and task rotation (NIOSH 1997, 34).

This discussion focuses on the systems being developed and not on the behavior of individuals. Safe versus unsafe ergonomic behavior is an entirely separate discussion from this section's discussion of recognizing a system's efficiency and efficacy by means of metrics.

Ergonomics training can take various forms but normally includes recognition of MSD hazards as well as discussion of ways to eliminate these hazards (NIOSH 1997, 13). The training can be given to various groups—all employees, only safety and ergonomics committee members, or members of management only. Can metrics be formed around ergonomics training? Yes—if the purpose of training is kept in mind. All training is conducted to impart knowledge (developing skills or competencies) or change behavior (NIOSH 1999, 5). It is easy to set simple metrics for training. One metric might require that 100 percent of a facility's personnel will have received ergonomics training within a one-year period.

Is this a fair measure of effectiveness? It is a leading metric, but its predictive value is limited and years are needed to recognize whether the training has affected the occurrence of injuries. Although training may have also been provided, no system was established to monitor the implementation of new knowledge or skills. Training must tie into other efforts. Once the purpose of training is clear, it can be tied into the information or skill being taught. If training is being conducted for the purpose of task analysis, make the training requirement part of the task-analysis metric. If the training is part of a reduction effort, tie it into that; make ergonomic improvement suggestions. A metric could read, for example, as follows:

- The facility will train 100 percent of its associates in ergonomic principles and the ergonomic suggestion process during the course of one year, collecting, tracking, and analyzing all suggestions.
- By the second year, the facility will have addressed within 30 days each ergonomic suggestion received and will have responded with its findings. If the facility has not received at least 25 viable suggestions by the end of the second year, the program will be reevaluated for continuation.

A metric developed for training requires an understanding of what the training should accomplish. As above, monitors can be built directly into the metric.

How can metrics be formed for other administrative items, such as forming ergonomics committees? A number of options are available; the simplest is to set a time limit, as in the following metric:

- During the first three months of the year, the facility will select the committee's members.
- During the second three months, the facility will provide the committee with training in the principles of ergonomics.
- By the year's end, the committee will have begun to generate corrective actions.

This is a very simple example, and its effectiveness is questionable. Not only is no monitoring component included, but no means exists for discovering whether the committee's efforts are affecting the rate of ergonomic incidents. The formation of the committee is the ultimate measurable outcome, but nothing is noted about how the incident rate might be lowered. What actions will the committee take? What actions will it particularly emphasize? Ergonomic hazards must be identified, corrective actions implemented, and the effectiveness of corrective actions monitored. The committee can assist in the accomplishment of all these functions. The following example incorporates them all:

- During the first quarter of the year, members of the ergonomics committee will be chosen and its first meeting will be held. During this meeting, the mission of the committee will be discussed and the members will draft suggestions for the committee's method of functioning.
- During the second quarter, the committee members will be trained how to recognize ergonomic hazards and how to develop ergonomic corrective actions. The committee will then decide upon a mission statement and brainstorm appropriate ways for the committee to:
  – Track symptoms
  – Recognize hazards
  – Suggest corrective actions
  – Follow up on implemented corrective actions
- During the third quarter, the committee will have communicated to management their finalized mission statement and finalized recommendations on the actions listed above. Management will respond with approval or alternative suggestions within the stated timeframe.
- During the last quarter of the year, the committee will have met and will be operating according to its mission statement and stated procedures.

This metric for forming the committee has more structure than the previous example, but it still needs to measure and monitor effectiveness. This can be fixed by adding the following:

- During the third quarter, the committee members will determine goals for themselves for the next years regarding all of the committee elements, also developing a monitoring system for symptoms that will be regularly updated. For every corrective action implemented, the committee will provide a write-up describing the former procedure or work area setup, what was changed, and what the effects of implementation were. Because of the close correlation of ergonomics with efficiency, these effects may include piece rate, quality levels, comfort rating, or other such things.

This provides the committee with an opportunity to develop its own metric while giving the company an avenue by which to monitor the committee's improvements to the overall ergonomic processes (by monitoring symptoms and related items).

The final administrative function to look at is task rotation. Task rotation decreases the duration of exposure to hazards, increasing the time away (*healing time*). Employees are rotated from tasks having ergonomic hazards to those having no (or less, or different) ergonomic hazards. This procedure, similar to engineering principles, is highly structured. The confounding factors are task complexities and the skill sets required. In facilities having very structured job descriptions, task rotation may be a viable solution. In facilities having high levels of job differentiation—especially in tasks that rely on employee experience and knowledge—task rotation may not be a good solution to ergonomic issues. Thus, one of the first issues in metric formation is deciding if task rotation is an ergonomic answer for a specific facility. Another factor that complicates task-rotation metric formation is training. How can employees rotate to jobs they are not trained in? Often training will be part of an initial metric. An example of a task-rotation metric might read as follows:

- During the first six months of the year, the facility will conduct a hazards analysis for ergonomic hazards and identify tasks posing a high number of risk factors.
- During the second six months of the year, the facility will identify jobs appropriate for job rotation, cross-train employees to be able to perform such tasks, and create a rotation schedule.
- The second-year job rotation will be monitored, each department reporting the percentage of compliance with the rotation schedule.
- After the third-year job rotation, the program will be reevaluated and a recommendation made about its continuation.

As in other cases, monitoring symptoms provides enough data to evaluate the effectiveness of such a program. The metric above demonstrates that even program elements of more subjective natures can serve as the bases for effective metrics.

Audit Monitoring

The audit can use a standardized corporate audit format or even be a simple *walk-through* audit.

If a company already has a structured, defined audit protocol, ergonomic elements can be incorporated into it, from hazard identification to corrective measures—the gamut of ergonomic processes. If the company sets goals and metrics around audit scores for a facility, the ergonomic elements could be incorporated in that fashion. A company could develop an ergonomic audit protocol and then set metrics for scores on such an audit.

Even walk-throughs can have metrics attached to them. Metrics can deal with everything from repeat findings (requiring that a department will not have a repeat finding within a one-year period) to corrective actions (requiring that any ergonomic hazards noted must have a corrective action plan developed within 30 days). As a method for monitoring the safety process, auditing offers an easy basis for metrics, including ergonomic metrics.

BENCHMARKING ACROSS COMPANIES

The majority of this chapter has outlined how an organization can benchmark against itself. It is the author's opinion that, due to the complexities of ergonomic injuries, the variation in how tasks are performed between (as well as within) different organizations and the difficulties in assuring standardization of programs, the various confounding factors (previously stated) are best controlled within an organization. There are benefits to be had, however, by comparing across organizations. These benefits include:

- greater diversity of programs
- larger exposure group
- greater number of data points

When comparing data across organizations, the difficulties often outweigh the benefits.

Benchmarking can be performed across organizations if a set of standards is imposed, and the variables between companies are minimized. There is a type of organization where this occurs naturally: the trade association—specifically, the vertical industry trade association. In a vertical industry trade association, all of the members are from a specific industry (paper, injection plastics, commercial banking, and so on). However, the differences between the members of a horizontal trade association may be too great for effective comparison.

Take the National Association of Manufacturers (NAM) as an example. Manufacturing can encompass a large variety of organizations, so large that standardization will be difficult. Possibly a medical definition of an MSD can be agreed upon and tracked for this diverse group, begging the question: What is the value of this metric? Different types of manufacturing will have different MSD rates. The association also does not have a leading component to help with predictability.

In a vertical trade organization, the members will have very similar operations. These similar operations can mirror the situation found within a single company. These similarities make benchmarking possible. The author was a member of the health and safety committee of the American Forestry and Paper Association (AF&PA), as well as the ergonomic subcommittee of that group. Such a group could decide upon and implement a specific ergonomic improvement, and the members could report back and benchmark against each other. The benefits would be the same as if they were one company.

Conclusion

Many analytical approaches can be used to evaluate ergonomic hazards. These analytical approaches can (and should) be treated as any other tool. Once the reasoning behind goal and metric formulation is understood, safety professionals can decide what to measure and decide what criteria will be established. Ergonomics can be measured in the same way as any other process. Although the examples used here will likely not work for most facilities, the principles they exhibit can be the basis for a facility's, company's, or corporation's development of metrics that will work for it. When the program and leading metrics are harmonized, the outcome will focus most on lagging metrics, or fewer injuries—the ultimate goal.

Summary

Forming metrics for ergonomics programs is a worthwhile endeavor. It allows safety professionals to evaluate the relevance of various components of such a program, assessing whether resources are being applied efficiently and outlining both immediate and (likely) future benefits. By following the general outlines offered in this chapter (that is, by measuring an essential element of a program's success over time), safety professionals can benefit from the formation of metrics and quantifiably assess their programs.

Endnote

[1] Prior to the mandatory OSHA record-keeping standard, voluntary reporting was conducted by organizations such as the National Safety Council. The membership of the voluntary reporting organizations did not include industrywide representation.

References

Bureau of Labor Statistics (BLS). 2006. *Injuries, Illnesses and Fatalities* (retrieved February 6, 2008). www.bls.gov/iif

California Department of Industrial Relations. *DWC Glossary of Workers' Compensation Terms for Injured*

Workers (retrieved April 15, 2007). www.dir.ca.gov/dwc/WCGlossary.htm

Heath and Safety Executive (HSE). 1999. *Health and Safety Benchmarking: Improving Together*. Sudbury, Suffolk, UK: HSE.

Heinrich, W. H. 1931. *Industrial Accident Prevention*. New York; London: McGraw Hill Book Company.

Matheson Discussion Group, The. u.d. *The Matheson System: The Functional Evaluation System* (retrieved April 15, 2007). www.roymatheson.com/index.html

Nash, James. 2001. "Recordkeeping: OSHA Tries to Make It Simple." *Occupational Hazards Magazine* (October) 62(10).

National Institute for Occupational Safety and Health (NIOSH). 1994. *Application Manual for the Revised NIOSH Lifting Equation* (Publication 94-110). Cincinnati.: NIOSH

———. 1997. *Elements of Ergonomic Programs* (Publication 97-117). Cincinnati: NIOSH.

———. 1999. *A Model for Research on Training Effectiveness* (Publication 99-142). Cincinnati: NIOSH.

National Safety Council (NSC). 1997a. *Accident Prevention Manual for Business and Industry: Administrative & Programs*. 11th ed. Itasca, IL: NSC.

———. 1997b. *Accident Prevention Manual for Business and Industry: Engineering & Technology*. 11th ed. Itasca, IL: NSC.

———. 1997c. *Supervisors' Safety Manual*. 9th ed. Itasca, IL: NSC.

NHS Direct Online Health Encyclopedia. *Repetitive Strain Injury* (retrieved December 2006). www.nhsdirect.nhs.uk/en.asp?TopicID=389

Occupational Health and Safety Administration (OSHA). 1971. 29 CFR 1904, *Reporting of Fatality or Multiple Hospitalization Incidents*. Washington, D.C.: OSHA.

———. 2001 (January 1). *OSHA Revises Recordkeeping Regulations*. Washington, DC: OSHA.

———. 2002. *Job Hazard Analysis*. Washington, D.C.: OSHA.

———. 2005. *OSHA Forms for Recording Work-Related Injuries and Illness*. Washington, D.C.: OSHA.

———. 2006a. OSHA Compliance Directive 06-01 (CPL 02), *Site-Specific Targeting 2006 (SST-06)*. Washington, D.C.: OSHA.

———. 2006b. *Standard Interpretations: Recording an injury when physician recommends restriction but no restricted work is available* (retrieved April 30, 2006). www.osha.gov/pls/oshaweb/owadisp.show_document?p_table=INTERPRETATIONS&p_id=25435

Oregon OSHA (OR-OSHA). 2005 (November). *Easy Ergonomics: A Practical Approach to Improving the Workplace*. Salem, OR: OR-OSHA.

———. Undated. *Online Training Module Chapter 1—Management Commitment*. www.cbs.state.or.us/external/osha/educate/training/pages/100xml.html

Sanders, M. S., and E. J. McCormick. 1993. *Human Factors in Engineering and Design*. 7th ed. New York: McGraw Hill.

Stewart, J. M. 2002. *Managing for World Class Safety*. New York: John Wiley & Sons.

Streiner, David L. 2003. "Unicorns Do Exist: A Tutorial on "Proving" the Null Hypothesis." *Canadian Journal of Psychiatry* (December), pp. 756–761.

University of California at Berkeley, Mechanical Engineering Department. Undated. *Manual Crimping*. www.me/Berkeley.edu/ergo/casestudies/crimper.html

Webster's Ninth New Collegiate Dictionary. 1983. Springfield, MA: Merriam-Webster, Inc.

Wikipedia. 2007a. *Trend Analysis* (retrieved April 15, 2007). www.en.wikipedia.org/wiki/Trend_analysis

———. 2007b. *Statistics* (retrieved April 15, 2007). www.en.wikipedia.org/wiki/Statistics

Best Practices

Farhad Booeshaghi

6

LEARNING OBJECTIVES

- Understand the basic terminology of ergonomic hazards and repetitive strains (EHRS) best practices.
- Be able to apply a systematic approach to identify EHRS.
- Recognize which ergonomic tools to use for eliminating or reducing EHRS.
- Know which ergonomic tools to employ in the design of products and processes.

ACCORDING TO THE Liberty Mutual Workplace Safety Index (WSI) of 2009, the estimated amount of workplace injuries and illnesses in 2007 (Table 1) cost U.S. Workers' Compensation $52 billion dollars. This is an average expense of one billion dollars every week. The WSI gathers its data from insurance companies, such as Liberty Mutual, as well as the U.S. Bureau of Labor Statistics (BLS) and the National Academy of Social Insurance. These agencies and insurance companies release data regarding serious injuries that take place at any workplace (WSI 2009). In this chapter, an attempt is made to study and outline the fundamental concepts of ergonomic hazards and repetitive strains (EHRS) as they relate to the practices of various industries in eliminating or reducing repetitive motion injuries.

In the mid-1800s, Polish biologist Wojciech Jastrzebowski coined the term *ergonomics*, which means the study or science of work. Since World War II, many pioneers in the ergonomics field have refined this study and have referred to it as "fitting the task to the person." In fitting the task to the person, there are two types of workplace injuries to consider: (1) acute injuries, resulting from an accident; and (2) cumulative injuries, which develop over time. The science of ergonomics focuses mainly on cumulative injuries (Jastrzebowski 1857).

ERGONOMIC HAZARDS AND REPETITIVE STRAINS

A common workplace hazard is *repetitive strain*. Caused by repetitive motion, it is defined as "performing a task that requires the application or experience of force at a consistent frequency while holding a posture over time." Repetitive motions may become repetitive strains if the worker is exposed to undue physical stress, strain, or overexertion. *These conditions include vibration, noise, heat,*

TABLE 1

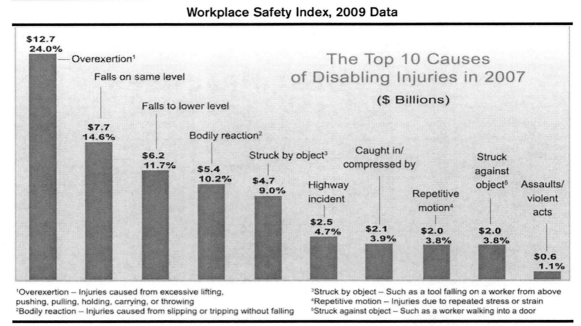

(*Source:* "The Most Disabling Workplace Injuries Cost Industry an Estimated $52 Billion," Liberty Mutual Research Institute for Safety, 2009.)

awkward postures, forceful exertions, contact stresses, and heavy manual material handling (Table 2). Repetitive strains result in cumulative injuries known as *musculoskeletal disorders* (MSDs). MSDs are injuries to the soft tissues—muscles, ligaments, tendons, joints, and cartilages—and the nervous system. Carpal tunnel syndrome, thoracic outlet syndrome, and tendonitis are examples of MSDs (Humantech 1996).

In the 1970s, these disorders began to appear in increasing numbers on companies' injury and illness logs. OSHA cited, and continues to cite, companies for hazardous workplace conditions that cause problems such as tendonitis, carpal tunnel syndrome, and back injuries. The Bureau of Labor Statistics (BLS), an agency of the U.S. Department of Labor, recognizes MSDs as serious workplace health hazards. In 2009 these injuries accounted for more than 40 percent of all lost-workday cases (BLS 2010).

Conditions that expose workers to repetitive strains are known as ergonomic hazards. Today's science of ergonomics looks at the cognitive or decision-making performance of humans as it relates to their daily interaction with their work environment, as well as the physical limitations of the interface, and attempts to eliminate or reduce ergonomic hazards. In best practices, engineers rely on the science of human-factors engineering and biomechanical engineering to identify, evaluate, and reduce or eliminate many workplace ergonomic hazards. Human factors that affect the mental, physical, and social behavior of workers and form their individual characteristics are functions of a plethora of variables, including but not limited to age, gender, size, strength, habits, literacy, experience, training, and physical and mental conditioning. A worker's perception-reaction, decision making, physical performance, and visual and auditory responses are examples of such human factors and are directly related to the potential of the worker being exposed to EHRS while performing his or her daily job. Elements of a job to consider include those shown in Table 3.

Human Factors Engineering

Although there are many opinions, it is for the aforementioned reason that many practicing engineers combine the term *ergonomics* with the term *human factors engineering* and refer to the combination as the *science*

TABLE 2

Sample Workplace Conditions that Expose Workers to Repetitive Strains

Conditions	Cause and Effect
Vibration	Vibration is an oscillation from the mean, regular periodic separation from the mean, which for a given application may become an ergonomic risk factor and affect blood flow, muscle contraction, and sensory systems.
Noise	Excessive exposure to auditory phenomenon from repetitive sound or static, due to unwanted perturbation of an audible sound, may become an ergonomic risk factor that can cause imbalanced inner-ear function.
Heat	The process of energy transfer from one medium to another, and the measure of this energy in terms of quantity, can become a probable cause for increased heat stress in employees who are actively working in hot conditions, such as the metal and rubber industries, road constructions, boiler rooms, bakeries, and so on.
Awkward postures	The need to maneuver the body to achieve an action in a workplace can be the primary cause of ergonomic risk. Examples include bending, lifting, and raising hands and shoulders to work on a surface that is higher than the body's torso.
Forceful exertions	Muscles exerted beyond their working limit fatigue and transfer exposed loading conditions to other supporting structures, such as ligaments and tissues. The loading conditions may exceed the nonload-bearing structure's capability and increase the worker's recovery period. This happens often if more than usual effort is attempted by using the limbs, tendons, ligaments, joints, and discs. Such extreme conditions create an ergonomic risk factor in the workplace.
Heavy manual material handling	MSDs are often related to manually handling materials that involve climbing, pushing, pulling, and pivoting. Since they involve injurious tasks, for a long and fixed period of time, these work processes have a high level of injuries associated with them.

(*Source:* Humantech 1996)

TABLE 3

Potential Worker Exposure to EHRS

Product	Process	User
Usage	Operation	Age
Usability of	Installation	Experience
Proper controls	Janitorial	Gender
	Service and maintenance	Stature
		Language/culture

(*Source:* Humantech 1996)

TABLE 4

Common Workplace Ergonomic Risk Factors

Body Part/ Work Action	Risk Factors
Hand and fingers	Pinch grip, power grip > 2, finger press
Wrist	Radial deviation, ulnar deviation, flexion > 45°, extension > 45°
Elbow	Forearm rotation, natural position 15° outward, inward rotation, outward rotation, full extension ≥ 135°
Shoulder	Raised > 45°, elbow behind body
Neck	Forward rotation > 20°, sideways rotation, backward rotation, twist
Back	Forward rotation > 20°, sideways rotation, backward rotation, twist
Leg	Straight stand, one-leg stand, squat, kneel
Force	Magnitude, direction, surface area, stress
Frequency (discrete or continuous)	Motion, noise, vibration, duration

(*Source:* Humantech 1996)

of ergonomics and human factors engineering. Biomechanical engineers study the kinetics (forces) and kinematics (associated path of motion) a person's body experiences in order to eliminate or reduce ergonomic hazards. Biomechanical engineering relies on bone mechanics, tissue engineering, and human anatomy and physiology to evaluate the exposure of workers to ergonomic hazards and repetitive strains. Issues associated with these hazards are ergonomic *risk factors*. Over time, ergonomic risk factors affect the human body and can cause permanent injuries. Table 4 presents some common workplace ergonomic risk factors.

Human factors engineering incorporated in the design of products and processes improves the quality of life for people in the workforce. Human factors engineering provides the means to identify design flaws, and/or the existence of risk factors in the occupational environment. These flaws and/or risk factors may limit workers' abilities and expose them to ergonomic hazards. Such exposure may affect workers in many different ways; for example, the regular use of a defective scissors may expose the hairstylist to the development of carpal tunnel syndrome (CTS). Adapting human factors engineering into product or process design as a system improves the reliability, maintainability, and usability of the system based on an optimum human performance. The purpose of this system is to provide improvements to the overall design, to relieve employees from stress and exhaustion, and to limit company liability exposure that may or may not occur due to end-user dissatisfaction. Well-established principles of human factors engineering are:

- Implement a flexible employee environment to accommodate workers' needs.
- Reassign hardware and software blueprints to limit an ineffective workplace environment.
- Assess and incorporate acceptable replication of human-related assignments.
- Suppress manpower errors through machine study and improved automation.
- Eliminate exposure to repetitive tasks through worker rotation programs.

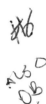

Incorporating these simple principles may reduce or eliminate workers' exposure to repetitive strains. In other words: *Strengthen the Strength and Complement the Weakness.*

Repetitive strains begin with discomfort, evolve into pain, and eventually become a cumulative set of disorders. Occupational safety and health professionals have called these disorders by a variety of names, including cumulative trauma disorders, cumulative trauma injuries, repeated trauma, repetitive stress disorders, repetitive stress injuries, and occupational overexertion syndrome. The injuries are painful and often disabling, and they develop gradually—over weeks, months, and years (Karwowski and Marras 1997; Adler, Goldoftas, and Levine 1997; Yassi 2000).

EHRS are routinely observed when humans and machines interface. Hand tools, workstations (seated or standing), manual material handling, platforms, ladders and stairs, controls and displays, the visual work environment, and machine clearances and maintenance are examples of human-machine interfaces (OSHA 2004).

EHRS are also observed in connection with these types of industries: construction (industrial/residential), medical and healthcare, nursing home, assisted care, child care, food service (restaurants/grocery stores), food processing and packaging, textile and fabric, mass storage and retrieval, manufacturing, retail and wholesale, transportation, postal service, service and repair, automotive and moving vehicle, household item, heating and air-conditioning, and cosmetology. They affect occupations such as custodians (janitors or cleaners), stock handlers and baggers, cashiers, carpenters, painters, electricians, roofers, cement masons, sheet metal workers, pavers, mail handlers, registered nurses, nurse's aides, orderlies and attendants, truck drivers, laborers, assemblers, typists, personal and office assistants, court reporters, barbers and hair stylists, teachers and lecturers, meat packers, ship builders, and more (Ernst, Koningsveld, and Van Der Molen 1997; Sitzman and Bloswick 2002; Melhorn 1998). The guidelines listed in this chapter may be applied to any industry or occupation that is not listed above and has similar occupational characteristics.

EHRS–Best Practices

Best practices for industries to reduce or eliminate EHRS rely on the application of scientific principles, methods, and data drawn from various disciplines, including biomechanics, industrial engineering, mechanical engineering, machine design, applied physical anthropometry, cognitive sciences, psychology, and physiology. The objective of best practices is to identify ergonomic risk factors associated with an industry, occupation, or task; to develop a best-practice ergonomic approach to eliminate or reduce those risk factors; and to outline specifications and criteria for the design of equipment, products, systems, and processes that are free from those factors.

Today, several industries use established control tools to eliminate or reduce EHRS effects on the human body. Examples of control tools include engineering controls that incorporate design, redesign, job-hazard analysis, work-site evaluation, human reliability analysis, strain index for reliability, and work-practices controls to manage ergonomic risk factors; administrative controls that include MSD management programs; training and education; risk-based programs; and PPE to guard against hazards.

The objective of this chapter is to outline some of the best practices industries use to eliminate or reduce the effects of EHRS on the human body. These guidelines are intended to provide the reader with the tools necessary to ensure that EHRS are addressed through each step in the product/process evaluation/design process. Since the fundamentals of EHRS best practices are the same among all industries, to accomplish this objective we begin by presenting the fundamentals of EHRS best practices followed by their applications in selected industries and occupations.

Fundamentals of Best Practices

As part of the elimination or reduction of EHRS in the workplace we begin by trying to understand human-machine interface requirements. When a product or a process is being designed or modified, the first step in eliminating EHRS risks is identifying the risk factors associated with its worker interface. Examples of products or processes with human-machine interfaces are workstations (seated, standing, or seated/standing), hand tools; controls and displays; manual material handling; machine clearances and maintenance; and platforms, ladders, and steps. The next step is to design or redesign the product or the process to eliminate or reduce EHRS. To accomplish this, *human-machine task allocation* (HMTA) must be studied.

HMTA focuses on human strengths, such as decision making, thought processing, and teamwork. It also takes into account the machine's characteristics—such as repetition, force, and operation—and the work environment, which includes thermal, auditory, visual, chemical, and physical effects. To design or modify the worker's interface with the process or product, it is also important to consider the worker's gender, body size, age, training for proper tool use, and posture, as well as the direction of tool and trigger travel (e.g., pull versus push).

HAND TOOLS

The need for greater productivity and less-expensive products and processes has mandated a more efficient workplace, resulting in safer hand tools. In the last decade, in order to reduce the problems associated with EHRS and simultaneously increase tool efficiency, improved hand tools have been designed and developed. These tools are often called *ergonomic tools*. Ergonomic hand tools such as wrenches, ratchets and sockets, screwdrivers, hammers, pliers, scissors, plumbing tools, punches, chisels, pry bars, gardening tools, cutters, files, and knives are now available. Most industries have adopted the concept of choosing the right tool for the job and using it correctly to avoid EHRS and potentially permanent and serious injuries.

It is impractical to identify one proper hand tool for every job or for every user. For example, a hammer with a large-diameter handle may decrease grip strength for a worker with a large hand, whereas it is unusable for a worker with a small hand. Each person has his or her own optimum grip diameter. The average person's optimum grip diameter, for a cylindrically shaped handle, is approximately 1.5 inches (Radwin and Haney 1996). The fit of a tool's handle, in terms of comfort and ease of use, is particularly important for tools used in long-term occupational activities, such as a carpenter's hammer or a drill. A fitted tool handle reduces the magnitude of concentrated forces experienced by the worker's hand during the tool's normal usage. It also reduces the amount of gripping force needed to cradle and oppose the reactive forces generated when the tool is used.

OSHA Guidelines

According to the Occupational Safety and Health Administration (OSHA), the following pointers should be kept in mind when selecting tools for use. These pointers apply to frequently used tools for kitchen work, gardening, housekeeping, laundry, and other maintenance areas, as well as those used daily at an operator's job, such as carpentry or plumbing tools (OSHA 2002):

- fits the job being done
- fits the work space available
- reduces the force one needs to apply
- fits the worker's hand
- can be used in a comfortable work position

Other guidelines for tool usage include using bent-handled tools to avoid wrist injury, using tools of an appropriate weight, selecting tools that have minimal vibration or utilize vibration-damping devices, maintaining tools regularly, and wearing appropriate personal protective equipment.

Hand-Tool Design and Selection Guidelines

Variations in hand-tool type, usage, and performance, coupled with variations in human anthropometry, make hand-tool design and selection challenging. Hence, understanding the general ergonomic risk factors associated with hand tools and adapting them to specific tools become essential parts of the hand-tool design or selection process (Kroemer et al. 2002). In general, the ergonomic risk factors for hand tools include:

1. awkward or forceful gripping of a tool
2. awkward body posture
3. holding/supporting a tool's weight
4. repetitive motion
5. vibration transmitted to a person's body, including the hand and wrist
6. high noise-level exposure
7. poor lighting conditions
8. heat/cold exposure
9. the operator's characteristics, including gender, size, knowledge, skill, and training

To create a hand tool with ergonomic features, a designer must focus on eliminating or reducing these risk factors. For example, tools should be designed, modified, or used in a manner that allows the forearm to be in a near-neutral position, 15° from pronation (the angle that the forearm makes with the horizon when the palm is facing down). Heavy tools should be suspended from above so the bulk of the weight is not supported by the worker. Tool handles should extend the full length of the palm and be soft, shock-resistant, and large enough to grip easily. Trigger-activated tools should be modified to allow for multi-finger operation, which prevents the required activation force from being applied by only one finger (Humantech 1996).

Several common materials can be used to modify tools, especially the tool-handle area, to improve hand grip and reduce vibration. Some hand tools are also designed for ease of one-handed use.

Awkward or Forceful Gripping

The ability to apply grip strength to a tool depends upon the tool-handle type, size, and length. In general, tools with longer handles provide more contact surface and require less strength, allowing the user to generate more leverage by applying a smaller force at a greater distance. A thick tool handle provides a greater surface for grasping and less stress on the hand. The stress is directly related to the amount of concentrated force applied over a small area on the hand. As either the applied concentrated force increases or the area where the force is applied decreases, the stress experienced by the tissues, muscles, tendons, ligaments, and bones in the hand increases. Repetitive exposure to high stresses may result in the loss of mechanical load-bearing capabilities for the human elements. For example, if a man routinely uses a pair of pliers with a 2.5-inch handle to remove rusted nails from boards, and the span of his hand is 3.5 inches, the bottom of the tool's handle section will exert a repetitive concentrated point force on the center of his palm. Over time, the stresses experienced may result in the loss of physical and mechanical characteristics of his hand. The main idea of redesigning the tool is to avoid concentrated pressure on small parts of the fingers or on the palm of the hand. For instance, plumbers add extensions to their pipe wrenches when loosening rusted pipes and use screwdrivers with thicker handles to generate higher torque, thereby reducing the overall required force. However, bigger and thicker handles do not always indicate an ergonomically safe tool. The user's ability, the tool's frequency of use and its application, and the allowable working space also affect the sizing of the handle. Changing the handle can reduce the force or grip strength necessary to use a tool. Even though larger and thicker tool handles facilitate application of grip force and reduce stresses on the palm, consideration should be given to the upper limit of the grip-handle design. An overextended tool handle may in itself become a problem and create new hazards and discomforts, such as catching on surrounding items or being too big for the available work space. Also, fine work may not be performed as efficiently with larger tools.

An optimum cylindrical gripping diameter is approximately 1.5 inches. The "OK" method is one way to determine the optimum grip diameter for an individual. In this method, the perimeter made by the thumb and index finger when making an "OK" sign indicates the optimum gripping diameter (Radwin and Haney 1996).

The best ergonomic designs for tools, based on the repetitive motions involved in their use, are those that avoid form-fitting contoured handles that fit only one hand size (Smith 2004). Today's ergonomically designed tools are made with slightly wider handles, which distribute the grip force over a large surface and thus decrease the necessary grip strength and resultant contact stress. Ergonomically designed tools that require opening and closing permit use by people wearing gloves, people with both small and large hands, and people who are either left- or right-handed. Hand tools with cushion grips provide comfort, slip resistance, and reduced grip force. A handle flange or handle taper is also used to reduce required grip strength.

Both variation in human anatomy and the way a tool will be used necessitate customization for a person and/or a job. A variety of materials can be used to customize or modify the handles of most hand tools to enhance their ability to be gripped. Commercial materials include Monoprene® thermoplastic elastomers, thermoset rubber, thermoplastic urethanes, Magic Wrap®, Plasti Dip®, GripStrip®, tool wraps, and pipe insulation. Several of these materials are readily available in most hardware stores. Gloves with slip-resistant material on the palm and fingers are also available. In addition to making tool handles thicker, materials such as thermoset rubber or epoxy putty can be applied to create custom molded finger grips. A wraparound handle allows the tool to stay on the hand with minimal effort, and a handle guard may be added to certain tools to prevent the hand from slipping forward onto the blade.

HANDLE TYPES AND SIZES

Hand tools that require two handles, such as pliers, generally have a 4.5-inch minimum handle length and a 3.5-inch maximum handle span. These tools are typically spring loaded to allow for easy separation of the handles. Spring-loaded handles eliminate the thumb force required to open the blades for sequential cuts and minimize hand fatigue (Kroemer et al. 2002).

Hand tools with one handle, such as drills, are equipped with either an in-line handle or a pistol-grip handle. In tools with in-line handles, such as screwdrivers, the tool action is in line with the grip handle. In tools with pistol-grip handles, the tool action is at an angle to the handle, typically 100°. In-line handles should be designed with a surface material such as rubber or closed-cell foam that has a high coefficient of friction, is compressible, and absorbs energy or vibration. However, materials that absorb moisture should not be used in a handle as they attract bacteria, can make the tool heavier, and can cause slippage by absorbing oils (Humantech 1996).

Pistol-grip handles provide greater torque while requiring less grip force than in-line handles. These handles should have a flange to prevent the fingers from slipping off, but the tool handle should not have finger grooves, since they cannot be designed to fit users' hands universally. Grooves also separate the fingers, causing localized pressure and stress concentration in the knuckles, joints, and finger ligaments, and they decrease grip force. Most pistol-grip tool handles are 1.5 inches to 2.5 inches wide (front to back) and the angle between the handle and the tool bit is around 100° (Putz-Anderson 1998).

A majority of triggers on pistol-grip tools are at least 1.5 inches long to allow for two-fingered triggering, using the index and middle fingers. Most three- or four-finger triggers are limited to use on suspended tooling systems. A good number of tools require four pounds or less static force to activate their triggers. Trigger strips or pressure-activated triggers are used on balanced or suspended tools. Thumb triggers are used in repetitive operations since our thumbs have greater capacity for limited action than our index or middle fingers, which are more prone to overexertion and injury. Lever triggers, which are activated by the palm or the inside of the thumb, are a minimum of 4 inches long and require four pounds or less to activate. They also have a safety catch to prevent unintentional activation. Pistol-grip power tools with a torque output exceeding 24 inch-pounds are equipped with a torque limiter to eliminate reactive forces.

Circular handles, averaging 1.5-inches in diameter, are used for power-grip applications. For tools used in precision operations, handles average 0.3 to 0.6 inches in diameter. In-line tool handles have oval grips with cross-sectional measurements of 1.25 to 1.75 inches. Straight or in-line power tools with a torque exceeding 14 inch-pounds are equipped with a torque limiter to reduce reactive forces. Slip-clutch, torque shut-off, hydraulic pulse, and torque reaction bars on balancers are examples of torque-limiting devices. Table 5 illustrates the sizing requirements for various kinds of hand tools (Helander 1995).

TOOL MAINTENANCE

Regularly scheduled maintenance of tools, including lubrication and replacing parts, keeps them at optimum performance and greatly reduces the forces exerted on the operator during use. For example, a worn drill bit requires more force to use than a sharp one; rusted pliers require more force to open and close than those with lubrication; and a broken saw blade requires a tighter grip to prevent the saw from kickback than a complete one. Saw blades that are Teflon coated or coated with other nonstick materials improve tool efficiency and reduce the forces the worker must apply. Sharp tools, maintained according to the manufacturer's specifications, also reduce the forces the worker must apply and thus reduce the possibility of injury (see Table 4).

Awkward Body Postures

Completing the task at hand, even given the proper tool for the job, sometimes requires an awkward body position. Posture greatly affects the forces experienced

TABLE 5

Handle Types and Sizing Requirements

Handle Type	Required Handle Length	Required Handle Diameter (Span)
Two handles (pliers, etc.)	4.5 in.–5.5 in.	0.5 in.–1.0 in. (2.5 in.–3.5 in.)
One handle in-line	4.0 in.–5.0 in.	1.25 in.–1.75 in.
One handle pistol-grip	5.5 in.–6.5 in.	1.5 in.–2.5 in.
Circular handle power grip	4.0 in. minimum	1.5 in.
Circular handle precision	2.5 in. minimum	0.3 in.–0.6 in.

(*Source:* Humantech, 1996)

by body parts during a tool's usage. Affected areas include the hands, wrists, elbows, shoulders, neck, back, hips, knees, ankles, and feet. Whenever possible, tools should be designed so that the user can occupy a more neutral posture. Tools can also be fabricated so that the user is forced to employ a less injurious position during their use.

Posture risk factors, associated specifically with the hands and wrists, include pinch grip, finger press, radial deviation, ulnar deviation, flexion, and extension. Poor wrist positioning can also diminish grip strength. For instance, a study has shown that grip strength is decreased by 27 percent when a wrist is held in flexion, 23 percent in extension, 17 percent in radial deviation, and 14 percent in ulnar deviation (Terrel and Purswell 1976). Poor wrist positioning can also lead to repetitive strain injuries. It is best to use hand tools that minimize flexion, extension, and deviation. Several hammers and pliers are designed with a bent or curved handle to maintain a more neutral wrist position, and some tools, such as gardening tools or paintbrushes, can be modified with an add-on pistol grip that allows for a more neutral wrist position.

Arm extension and forearm rotation can be factors that affect the elbows. Elbows should be kept in front of and as close to the body as possible during a hand tool's usage in order to eliminate risk factors affecting the shoulders. For instance, ergonomically designed computer keyboards are set at 15° to eliminate forearm rotation. In addition, the posture required when using a hand tool should minimize or eliminate bending or rotating the neck forward, backward, or sideways. Hand-tool operations that require the operator to squat, stand on one leg, or kneel should be modified to eliminate the forces generated on the hips, knees, ankles, and feet.

In studies of workplace EHRS, *best work zone* and *preferred work zone* are commonly identified for operators performing a given task. These zones help to eliminate ergonomic risk factors and are defined in the following ways: the *best work zone* for an operator using a tool is between shoulder width, in front of the body, above the hip, below shoulder height, with the elbow next to the body; the *preferred work zone* is the region created by the rotation of the arms about the shoulder, in front of the body, and below shoulder height (Pulat 1997).

Obviously, it is best to stay within the best work zone or the preferred work zone when using a hand tool; however, this may not always be possible. Hence, pistol-grip and angled-handle tools should be used on vertical surfaces that are at elbow height and on horizontal surfaces that are below waist height. Straight or in-line tools should be used on horizontal surfaces that are elbow height and on vertical surfaces that are below waist height. For example, knives with angled and pistol-grip handles are designed for cuts made with a downward stroke. These knives may not be widely used; however, they keep the wrist in a neutral position while allowing for sufficient downward force to make a smooth cut.

Holding or Supporting a Tool's Weight

Hand tools that require substantial strength to support are enhanced by external means. Heavy tools are equipped with two handles and are suspended or counterbalanced. Even if a user can comfortably support the tool's weight for a short period of time, over a longer period, the weight will statically load the user's muscles and diminish ability to support the tool. To compensate for the muscle's diminishing ability to carry the tool's weight, forces are transmitted to the joints, tissues, and ligaments, exposing the body to ergonomic risk factors that result in over-straining. Tool enhancements such as spring balancers, retractors, line reel balancers, Ergo-Arms® (articulated joints used in manufacturing lines to support the weight of a tool), and air-hose support systems are used to eliminate or reduce static muscle loading associated with tool weight.

Repetitive Motion

The single largest occupational health hazard in the United States is an injury sustained by excessive use or abuse of muscles, tendons, and nerves, resulting in *repetitive motion injuries*, also known as *repetitive strain injuries* (BLS 2010).

Highly repetitive tool use, which includes short-cycle motions repeated continuously over a long period of time, may eventually result in EHRS injuries. Elimination or reduction of the repetitive motions necessary to complete a task often requires a redesign of the entire process (Smith 2004). For example, if a repetitive assembly task has sufficient clearance, changing to tools with a ratcheting mechanism or gears can help reduce repetition. Maintaining hand tools properly (e.g., keeping saw blades and drill bits sharpened) and using proper operating methods (e.g., making pilot holes for drilling) can also reduce the required grip force and therefore reduce repetition. If the work environment allows for it, changing to a power tool might also reduce repetitive motions. If possible, switching to hand tools that have adjustable spring-loaded returns, such as pliers and scissors that open automatically, can reduce repetition. And finally, some innovative hand tools can also reduce repetitive motions because of their design; for example, the blade of the Stanley SharpTooth Tool Box Saw® reportedly cuts 50 percent faster than a standard hand saw due to a unique tooth design that cuts in both directions (DoItYourself.com 2007).

If it is not possible to reduce the repetitive motions necessary to use a hand tool due to the nature of the job/task, it may be beneficial to plan or redesign the task itself. Sometimes a few hours of employee brainstorming and problem solving may not only increase morale, but may result in solutions that save time and costly medical bills later.

Vibration

Vibration in tools is generally associated with power hand tools that use sources of power such as air, electricity, and gas. Typically, power hand tools are used when greater applied force is required, repetitive tasks are being performed, or time can be saved. With the advantages of power tools also may come some disadvantages, including vibration, repetitive strains such as trigger finger, and increased operator demands and requirements to handle and react to the forces generated by the power tool.

Among these disadvantages, vibration may pose the greatest concern. Exposure to large amounts of vibration in a localized area, such as the hand, over a prolonged period of time, can increase the risk of chronic disorders of the muscles, nerves, and tendons. Some studies have shown vibration to cause temporary sensory impairments (Streeter 1970, Radwin et al. 1990).

Although vibration is sometimes a desired effect, as with sanders and grinders, most often it is an undesirable by-product of power-tool use. Existing hand-tool guidelines tend to focus on areas where the effects on humans are measurable—for example, the amount of vibration transmitted (ISO 5349-1986, Acoustical Society of America ANSI S.270-2006). The amount of vibration transmitted by a power tool can be influenced by a tool's weight, design, and attachments (such as a power line). Proper maintenance of power tools is a top priority in order to prevent added vibration due to a failing bearing or worn, out-of-balance parts. Tools that produce low-frequency/high-amplitude vibration, that is, 8–1000 Hz, are considered high risk.

Power tools designed with anti-vibration materials or anti-vibration mounts or handles have had limited success in reducing the amount of transmitted vibration. If the job or work environment requires the operator to use power tools for prolonged periods, it is best to consider redesigning the process, redistributing the work, or using some kind of external support for the power tool. Gloves made with materials that dampen vibrations transmitted to the hands, wrists, and arms may be worn, but their effectiveness may vary.

High Noise-Level Exposure

Excessive noise in the workplace results in operator exposure to ergonomic risk factors. Hearing loss, communication interference, annoying distractions, and tuning out are examples of such risk factors. In addition to the ergonomic challenges they present, these risk factors may lead to a decrease in productivity. The intensity, frequency, and duration of noise determine the level of risk. Exposure to a high noise level affects people gradually, beginning with reduction in quality and clarity of high-frequency sounds. Noise-level exposure in excess of 90 dBA across an 8-hour period is considered an ergonomic hazard. Where noise levels are excessive, administrative or engineering control tools should be integrated into the work process to minimize the noise associated with any ergonomic risk factors. Requiring operators to wear hearing protection, shortening operators' exposure time to high noise levels, and educating and training the operators are examples of administrative controls. Using mufflers to reduce the noise from the release of air or exhaust from tools, isolating sound, and automating high noise-level tasks are examples of engineering controls. Eliminating the source of noise is far more beneficial than guarding against it.

Examples of ways to eliminate noise include:

- replacing old machinery with newer more robust equipment
- lubricating moving machinery parts
- fitting the pneumatic tool's intake and exhaust system with mufflers
- modifying the material-handling process to eliminate the impact or sliding of metal on metal
- controlling resonance

Poor Lighting Conditions

Poor lighting conditions that result from illumination, glare, and color can affect the operator's vision, performance, and safety. Additionally, inadequate illumination can lead to eye strain and headache. The amount of luminance needed is a function of the task at hand, the duration of the task, the reflectance of the environment, and the operator's visual capability. Typically, light levels are measured using a photometer, a device used to determine light levels in different planes. Light can be quantified in many ways—lux, lumens, footcandles (fc), candlepower, candelas, and so on. The two most popular scales are lux, the European measure, and footcandles, the U.S. measure. Lux equals the illumination of one square meter one meter away from a uniform light source. It is also defined as one lumen per square meter. One candela is equal to one lux. A footcandle is the illumination of one square foot one foot away from a uniform light source or one lumen per square foot. One footcandle is approximately ten lux (1 fc = 10.76 lx).Typically, a common area with dark surroundings is around 50 lux when illuminated. A workplace where fine work is performed over a long period should be about 5000 lux.

Quality of illumination can often be improved by reducing glare. Eliminate direct glare, where light directly affects the vision—such as sun in the eye. Surfaces that diffuse light, such as flat paint, matte paper, and textured finishes, are preferred. If the direct glare cannot be eliminated, use light shields, hoods, and visors to guard against it. If possible, change the orientation of the workplace, task, viewing angle, or viewing direction until maximum visibility is achieved. Use of color determines a person's perception of the work environment. Areas covered with long-wavelength colors, such as red, appear smaller than areas covered with shorter-wavelength colors, such as blue or green. Color brightness affects an operator's perception. Dark colors appear closer than the lighter ones. Also, cool colors, such as blue, green, and violet, are soothing, whereas warm colors, such as red, orange, yellow, and brown, are stimulating.

Heat or Cold Exposure

Exposure to extreme temperatures can affect workers' job performance and expose them to ergonomic risk factors. For example, air-powered tools can decrease exposure to temperature extremes by cooling the tool handle or blowing out exhaust air. Metallic tool handles should be insulated with a rubber, plastic, or cloth covering that shields the hand from extreme temperatures. Skin exposure to direct air exhaust can be eliminated by using exhaust mufflers and by directing the exhaust away from the operator. OSHA recommends that the operator's skin should not be exposed to temperatures less than 65° Fahrenheit or greater than 95° for a long period of time. Tools used for joining—such as a soldering iron, a welding rod, a hot glue gun, or a sealant tube—produce heat energy. Face shields, goggles, and gloves eliminate the probability of exposure to such heat energies. Electric motors, friction surfaces, and exhaust mufflers are also sources of heat energy. Face shields, goggles, and gloves help reduce exposure to heat.

Operator Characteristics

The human characteristics of age, gender, ethnicity, stature, physical ability, health, personal habits, knowledge, training, education, and past experience are directly related to the potential ergonomic risk factors workers may experience while performing their jobs. A task that might expose one operator to an ergonomic risk factor may not have any effect on another who has slightly different physical or emotional characteristics. Determining operators' characteristics using examination and testing allows for the selection and design of a work environment that minimizes or eliminates exposure to potential ergonomic risk factors. A positive work environment, the promotion of physical health, and incentive programs that encourage workers to break bad habits can also reduce risk factors that are associated with an operator's characteristics.

Workstations

Employers need to decide whether it is appropriate for particular employees to be seated or standing while performing their jobs based on the tasks they perform regularly. Seated workstations are best when the job requires one or more of the following: precise foot control, fine assembly, writing or typing tasks, body stability or equilibrium, long work periods, or work less than 5.9 inches (15 centimeters) above the work surface.

An effective and ergonomic seated workstation will do the following: give the operator the correct eye position for viewing the task; allow the seat's height, depth, back angle, and footrest positions to be adjusted for comfortable working conditions; allow for clearance of the legs and knees; and allow adjustment of the work surface's height and depth. Guidelines for designing effective seated workstations are available in many textbooks and on specific industrial Web sites. OSHA provides a good Web site on this topic at www.rmis.rmfamily.com/new/ergon60.htm.

Standing workstations are best in the following situations: no leg or knee clearance is available; heavy objects are regularly handled; high, low, and extreme reach are frequently required to complete the task; and mobility is necessary. An effective standing workstation will allow proper eye position with respect to viewing requirements, permit proper reach distances, and allow the work surface height and depth to be adjusted (Lang 1999).

Manual Material Handling

Workplace injuries are most commonly associated with material handling. Most of these injuries are caused by heavy lifting. To guard against material-handling injuries, companies dealing with this type of work should advise their employees to follow these simple procedures:

- Always bend at the knees and not at the waist. Bending at the waist puts excessive strain on the lower back, which can result in serious back injury.
- When lifting a heavy object from a crouched position, do not jerk the body, particularly the back. Keep the back straight and lift with the legs.

Platforms, Ladders, and Stairs

Very few best practices exist for climbing stairs on a repetitive basis. Generally, it is recommended that workers maintain proper posture and not bounce. Proper posture is generally defined as keeping the back straight and lifting the weight of the body with the legs. To prevent the concentration of weight in one area of the body, a hand should be placed on the handrail at all times to brace the body and to allow for more leverage when climbing. Bouncing is common when running up stairs. This places extra stress on the joints and may lead to injury, so running should be avoided.

Examples of Best Practices by Industry

Gardening

Today's garden gloves feature high-tech materials and advanced designs that reduce hand fatigue, eliminate friction, and dampen vibrations, as well as protect the hand. Ironclad Performance Wear, which produces high-tech work gloves, suggests that the same performance offered in work gloves can be useful in gardening gloves (see Figure 1). Features such as a protected knuckle area; reinforced fingernail guards; washability; reinforced thumbs, saddles, and fingertips; breathability; adhesive grips; and seamless fingertips that increase the sense of touch make gardening easier and safer for the hands. Also, proper posture should be practiced when gardening in order to reduce common arm, hand, wrist,

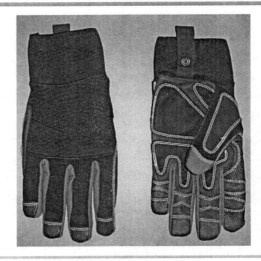

FIGURE 1. "Cargo Bull" by Ironclad® Performance Wear (*Source:* Ironclad Performance Wear, online at www.ironclad.com)

and back injuries. Maintaining a straight wrist can reduce some of these injuries. Bent-handle cutters and pliers help the user maintain a straight wrist.

Pharmaceuticals

Pharmaceutical technicians hand-tighten dozens of vaccine jug caps daily. If not adequately tightened, the jugs could leak and spoil products worth thousands of dollars. However, most operators are poor judges of cap torque and are marginally capable of using the proper torque required to tighten caps adequately. Caps that are not properly tightened may result in significant, unwarranted hand and wrist stress. One pharmaceutical company employs a dial torque wrench made with a special cap torque attachment and trains its technicians to use it. The cost is about $8 per worker.

Meatpacking

Meatpacking is one of the most hazardous industries in the United States because in assembly line processes, such as boning meats, workers can make several thousand repetitive motions per day with no variation. The motions place physical stress and strain on the wrists and hands, resulting in carpal tunnel syndrome. One company has automated material-handling and packaging processes to reduce or eliminate operators' exposure to ergonomic risk factors.

Garment Manufacturing

Garment makers, who often perform fast-paced piecework operations involving excessive repetitive tasks, increase their risk of developing carpal tunnel syndrome. Garment-industry jobs often require workers to push large amounts of materials through machinery while sitting on nonadjustable metal stools. Workers doing these jobs can sustain disabling wrist, back, and leg injuries. Automation as well as engineering and administrative controls are used to alleviate some of the risk factors. A garment-manufacturing workstation is shown in Figure 2.

Ceramic Cooktops

At a glass ceramic cooktop plant, workers manually lift uncut plates of glass onto a waist-high conveyor belt, where they are stacked vertically on a nearby L-shaped holder. A forklift moves the strapped holder carrying the glass. The holder, however, presented the glass at knee height, making workers bend each time to pick up the glass until they devised a stand made from a wooden shipping crate and placed it beneath the holder to raise the glass to waist height.

Professional Offices

Employees in many offices experience pain from performing their daily tasks. Workers should be trained in the proper use of adjustments already provided in their chairs, computer monitors, and furniture systems. Changes in the placement of telephones, printers, and in-boxes can lead to better working postures. In addition, training and encouraging employees to take micro-breaks helps overused parts of the body to rest

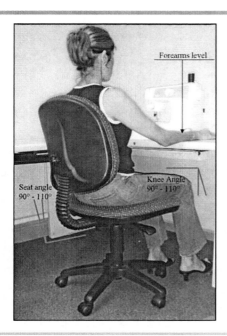

FIGURE 2. Garment-manufacturing workstation (*Source:* OSHA, www.osha.gov)

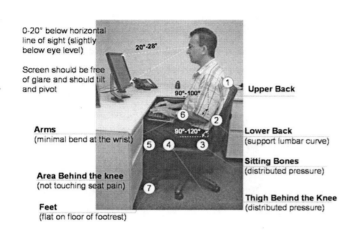

FIGURE 3. Professional office workstation
(*Source:* Nicholas Institute of Sports Medicine and Athletic Trauma, www.nismat.org)

and recuperate. A professional office workstation is shown in Figure 3.

Poultry Processing

Poultry processing involves many risk factors, including repetition, force, awkward and static postures, and vibration, all of which may contribute to an increased risk of injury. The type of work performed, in conjunction with cold temperatures that are associated with poultry processing, may also be a contributing factor for MSDs. Poultry processors have created a number of solutions to reduce the duration, frequency, and degree of exposure to risk factors. Developing a job-rotation system helps to reduce the fatigue and stress of particular sets of muscles and tendons. Cross-training and providing floating employees ensures that staff support is always available to cover worker breaks and provide rotation alternatives. The employees in one poultry-processing plant complained that ill-fitting protective gloves did not provide adequate protection, so the company purchased gloves from several manufacturers to provide a wider range of sizes for better fit. Also, operators' stationary seats were replaced by adjustable seats.

The Occupational Safety and Health Administration provides recommendations for the use of workstations that are appropriate for duties being performed (OSHA 2004).

Packaging

Workers used to pack items into rectangular boxes positioned so they had to reach repeatedly across the long axis of the boxes, exposing their backs, shoulders, and arms to physical stress. Rotating the boxes allowed them to reach across the shorter axis of the boxes, reducing the length of reach and the risk of injury. Manual lifting is often associated with the packaging industry. Excessively heavy boxes can place great stress on employees' muscles, leading to back strain or disc injury. Improving access to heavy boxes, allowing employees to pick them up without having to bend at the waist, and providing boxes with handles or handhold cutouts, can definitely reduce the risks of injury (access the OSHA Baggage Handling eTool).

Visual Display Terminals/Visual Work Environment/Computer Environment

Employees who spend much of their time in an office regularly stare at a computer screen or LCD display for large amounts of time. One common repetitive stress injury is damage to the eyes, often resulting in the need for glasses or contact lenses. It should be noted that flat-panel computer and workstation monitors emit much less radiation than some of the older model screens and are much better for the eyes.

Split keyboards are used in order to place less stress on wrists, hands, fingers, joints, and tendons when employees type for long periods. Much research has been done in the area of split-keyboard technology over the last few years. For example, Timothy Muss, a graduate student in the Human Factors and Ergonomics Laboratory at Cornell University, along with Alan Hedge, professor of Design and Environmental Analysis at Cornell, developed a vertical keyboard that reduces many of the stress-related injuries associated with long periods of typing. Using gloves with special sensors, the stresses placed on various areas of the wrists and hands when using the new keyboard were carefully measured and compared to the stresses generated when using a traditional keyboard. The vertical split keyboard, also known as the VK, was shown to be far superior to traditional keyboards in reducing repeated stress–related injuries (see Figure 4).

Machine Clearances and Maintenance

In the area of machine maintenance, best practices are generally specific to the job. Each machine has a different operating and maintenance procedure and its own best practices. Best practices common to all machines include easy access for maintenance and lubrication, clearances that allow for easy access to areas needing regularly scheduled maintenance, and engineering and administrative controls to reduce or eliminate workers' exposure to ergonomic risk factors while performing maintenance.

FIGURE 5. Fiberglass California Framer® (*Source:* www.hammernet.com/news.htm)

Construction–Industrial and Residential

In the construction industry, it is generally understood that workers use their own best practices to get jobs done fast. Most of the construction industry works on a strict timeline, and emphasis is on completing the job on time. An example of an ergonomic hammer, developed to reduce the continuous vibration that construction workers often endure, is the fiberglass California Framer®, pictured in Figure 5. This hammer, with its fiberglass handle and oversized striking face, reduces the stress that the wrist, hand, and arm often experience when hammering a nail.

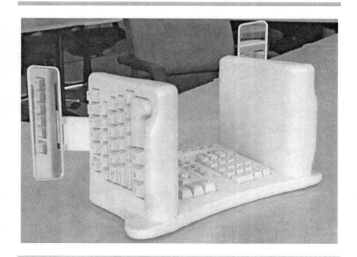

FIGURE 4. The VK, an ergonomically designed vertical keyboard (*Source:* www.news.cornell.edu)

Medical and Healthcare

There are many guidelines for proper procedures in all aspects of the medical and healthcare industry, most of which are controlled by OSHA. They include the proper ergonomic practices for repetitive motion that help prevent carpal tunnel syndrome. For example,

FIGURE 6. ErgoHold 3 scalpel handle
(*Source:* LM Instruments, www.lminstruments.com)

an ergonomic scalpel handle for surgeons was developed by LM Instruments OY. Known as the ErgoHold 3 (see Figure 6), it is an improvement over the traditional scalpel because it has a firm grip and good balance, and it allows for precise blade control as well as controlled rotation.

Nursing Homes, Assisted Care, and Child Care

Throughout the nursing-home, assisted-care, and child-care industries, there are no clear-cut ergonomics best practices. In 2003, OSHA published some basic guidelines intended to reduce ergonomic risk factors associated with these industries (OSHA 2003). The main focus of those guidelines was on lifting and repositioning patients. OSHA has recommended a number of solutions, some of which include the use of powered sit-to-stand devices (see Figures 7 a and b), portable lift devices, ceiling-mounted lift devices, variable-position geriatric and cardiac chairs, built-in or fixed bath lifts, and toilet-seat risers. This area could benefit from further research on reducing repetitive strain injuries in the workplace.

Food Processing and Packaging

There doesn't appear to ever have been a publicly released study on the best ergonomic box-cutter design in the industry. Several companies claim that their box cutter is the best and most ergonomic; such cutters are readily available in hardware stores. Ensuring that blades are sharp helps to reduce stress on the user. For an example, see Figure 8.

Textile and Fabric Manufacturing

In the textile and fabric industries, the use of precision cordless rotary shears reduces repetitive motion injuries due to cutting for long periods. One company instituted a program of monthly rotation of employees performing

FIGURE 7a. Sit-to-stand devices
(*Source:* OSHA, www.osha.gov)

FIGURE 7b. Sit-to-stand devices
(*Source:* OSHA, www.osha.gov)

FIGURE 8. Box cutters (*Source:* Safety Knife Company, www.safetyknife.net)

repetitive-motion jobs. They also designed equipment with the operator and task in mind, redefined processes to ensure employee safety, and educated employees about proper job procedures (BMP Center of Excellence for Best Manufacturing Processes 1994).

Service and Repair

Many products in the service and repair industry incorporate ergonomics best practices in their design. Two examples are fatigue-reducing needle-nose pliers by Klein Tools and MaxGrip™ self-adjusting pliers by Stanley. Comfort-grip pliers have spatulated handles molded from flexible, nonslip material for softness, comfort, and a secure grip. Spatulated handles expand contact with the hand, reducing pressure and strain and improving power transfer. Two types of comfort-grip pliers are shown in Figure 9.

High-Noise Work Environment

Some circumstances require employees to work in loud environments (> 90 dBA) where noise cannot be reduced any further, requiring ear protection based on OSHA standards (OSHA 2010). *Noise-canceling ear protection* (NCEP) is ideal in loud work environments that subject workers to prolonged exposure to high-decibel sounds. The NCEP technology has been around for the better part of 50 years, originally invented for

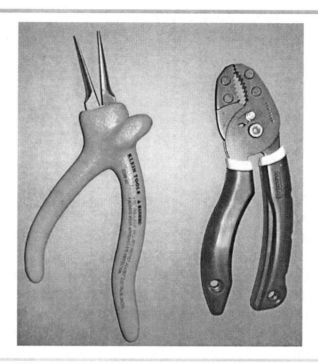

FIGURE 9. Klein fatigue-reducing pliers and Stanley MaxGrip™ self-adjusting pliers (*Source:* Grainger, www.grainger.com)

pilots to reduce engine noise and allow better radio communication in flight (Elliot 2001, 440–464). The technology responsible for functionality of the noise-canceling headphones is referred to as *active noise control*. Companies are now using active noise-control technology to improve ear protection. Today's noise-canceling ear protection has been designed to not only muffle sounds above the 90 dBA, ear-damaging level, but also pick up and amplify sounds around the 60 dBA level (OSHA 2010). NCEP works by mounting small microphones on each earphone; when the microphones detect noise, they create an antinoise signal in the earphone, causing destructive interference and eliminating unwanted ambient noise (Liu 2008). This allows workers to clearly communicate with one another in a loud work environment without damaging their hearing, which cannot be accomplished with earplugs. Noise-canceling ear protection has just recently become a more economical and viable option for employees in the workplace. Although noise-canceling ear protection is not as affordable as earplugs, it is found to be more comfortable and ergonomic for industries that require ear protection for their workers.

Cell/Mobile Phone Technology

During the last decade, cell phones, also known as mobile phones, have become the major communication tool for millions of employees in a multitude of industries. Hands-free devices for cell-phone users became a necessity when states began to prohibit talking on cell phones while driving, making the practice illegal. Bluetooth® technology prompted companies to develop hands-free headsets; used in conjunction with cell phones, they allow users to converse without the use of their hands (Bluetooth SIG 2011). In June of 2011, the World Health Organization (WHO) announced concerns about prolonged exposure to cell-phone electromagnetic radiation. The WHO report indicated that years of prolonged cell-phone use may cause the user to develop a brain tumor. Bluetooth headsets emit 1mW of power, compared to the 1W of power emitted by a cell phone, which means significantly less electromagnetic radiation exposure (Scarfone 2008). The practice of using a hands-free headset is safer than using a cell phone without one. In general, using cell phones while performing other activities can be dangerous for workers by causing distractions and impairing hearing while they work with heavy machinery or operate a motor vehicle.

Hydraulic Shock Absorption

Extensive research and development in the magnetorheological fluids (MRF) field is rapidly improving the ergonomics of shock absorbers and hydraulics (Rinaldi et al. 2004). MRF are non-Newtonian fluids; their viscosity is not constant, allowing for no linear relationship between shear stress and shear rate (Poynor n.d.). A shock absorber equipped with MRF is capable of varying the viscosity of the fluid by the use of an electric field. The application of an electric field causes the stiffness of the shocks to automatically adjust, and the damping characteristics are controlled, allowing a vehicle traversing rough terrain to perform better and give a smoother ride. Currently, MRF shock technology is being used in military applications such as Hummers (HMMWV). This technology allows the military Hummers to travel faster and safer than previously (LORD Corp. 2011). Large industrial vehicles that travel off road, such as logging and construction trucks, present a possible commercial application for MRF shock technology. It provides a smoother ride for the truck drivers, and it is an ergonomic upgrade; however, retrofitting vehicles with these systems would come at a hefty price.

Woodworking

The woodworking industry is known for incurring workplace injuries. Although there are many safety regulations to protect workers while handling and cutting wood, injuries still occur. In today's market, saw manufacturers are innovating safety shields and kill switches to enhance the ergonomics and safety of their products. One patented, preventive technology is designed to detect when a worker's body part contacts the saw blade, immediately stopping the blade and preventing injury to the worker. Table saws with this technology have blades that detect electrical signals while the saw is running (see Figure 10). When the saw blade comes into contact with an electrically conductive material, such as a human finger, a patented brake design stops the blade in less than a millisecond, preventing serious injury to the operator. This safety system works for saws designed to cut nonconductive materials such as wood (SawStop 2011).

FIGURE 10. Table saw with blade that detects electrical signals (*Source:* SawStop 2011)

Textile Screen Printing

Repetitive strain injuries are common in textile screen-printing industries, especially in large-scale production companies. Workers manually load garments onto automatic printing presses, later transferring the printed cloth into dryers that allow the ink to cure. Even though this task is not very strenuous, when workers print thousands of items a day, repetitive motion injuries are foreseeable. Automation has enhanced the process of loading and unloading shirts, towels, and other textiles, thus eliminating worker exposures to repetitive motions (M&R 2011). One such automation technology is known as a robo-pneumatic machine. While this machine can operate on its own, workers still have to keep a close eye on the loader to ensure that garments are properly centered on the pallets to produce a quality-aligned print. There is give and take with the automation, and altough the system may work effectively and prevent exposure to repetitive strain injuries, it may not perform as efficiently as a human operator.

High-Temperature Exposures

Infrared (IR) noncontact thermometers were designed to improve temperature measurements conducted in both materials-processing and food-manufacturing industries. By pointing a laser at the desired target, the device determines the temperature of the target by measuring the emitted blackbody radiation. IR thermometers allow workers in high-temperature, ceramic processing facilities to measure the temperature of a substrate from a safe distance, minimizing the risk of being burned. In the food-manufacturing industry, noncontact IR thermometers are used to minimize the workers' handling of the product. The temperature measurements are conducted without compromising the sanitation of the product. IR thermometers are limited in the range of temperature they can measure, affecting the accuracy of high- ($> 1370°C$) and low-temperature ($< -30°C$) measurements (Omega Engineering 2011).

IMPORTANT TERMS

Anthropometry: The study of human body measurements, used in developing design standards and requirements for manufactured products to ensure that they are suitable for their intended audience.

Biomechanics: A scientific and engineering field that explains the characteristics of the human body in mechanical terms.

Carpal Tunnel Syndrome: The compression and entrapment of the median nerve in the carpal tunnel, where it passes through the wrist into the hand. The median nerve is the main nerve that extends down the arm to the hand and provides the sense of touch in the thumb, index finger, middle finger, and half of the fourth, or ring, finger.

De Quervain's Disease: Inflammation of the tendon sheath of the thumb, attributed to excessive friction between two thumb tendons and their common sheath. It is usually caused by twisting and forceful gripping motions with the hands.

Extension: A position in which a joint is straightened so that the angle between the adjacent bones is increased.

Finger Press: The motion of the tip of a finger or fingers pressing down on an object, such as a button or key.

Flexion: A position in which a joint is bent so that the angle between the adjacent bones is decreased.

Industrial Hygiene: The science of anticipating, recognizing, evaluating, and controlling workplace conditions that may cause worker injuries and illnesses.

Kinesiology: The study of the principles of mechanics and anatomy in relation to human movement.

Musculoskeletal Disorders: Injuries and disorders of the soft tissues (muscles, tendons, ligaments, joints, and cartilage) and the nervous system.

Pinch Grip: A posture in which the hand does not fully encircle the handled object.

Radial Deviation: A posture in which the wrist is bent toward the thumb side of the hand.

Raynaud's Syndrome or White Finger: A condition in which blood vessels of the hand are damaged from repeated exposure to vibration for long periods of time. The skin and muscles do not get necessary oxygen from the blood and eventually die. Symptoms include: intermittent numbness and tingling in the fingers; pale, ashen, and cold skin; and the eventual loss of sensation and control in the hands and fingers.

Tendonitis: A tendon inflammation that occurs when a muscle or tendon is repeatedly tensed from overuse or unaccustomed use of the wrist and shoulder.

Tenosynovitis: An inflammation of or injury to the synovial sheath surrounding a tendon, usually due to excessive repetitive motion.

Trigger Finger: A tendon disorder that occurs when a groove is worn into the sheath of a flexing tendon of a finger, which is usually associated with using tools that have handles with hard or sharp edges. If the tendon becomes locked in the sheath, attempts to move the finger cause snapping and jerking movements.

Ulnar Deviation: A posture in which the wrist is bent toward the little finger side of the hand.

REFERENCES

Adler, Paul S., Barbara Goldoftas, and David I. Levine. 1997. "Ergonomics, Employee Involvement, and the Toyota Production System: A Case Study of NUMMI's 1993 Model Introduction." *Industrial and Labor Relations Review*, (April) 50(3):416–437.

Acoustical Society of America. 2006. ANSI S.270-2006, *Hand Arm Vibration Standard*. Melville, NY: Acoustical Society of America.

Bluetooth SIG. 2011. *Safety Benefits of Bluetooth Headsets* (retrieved June 21, 2011). www.bluetooth.com/Pages/Cell-Phone-Safety.aspx

BMP Center of Excellence for Best Manufacturing Processes. 1994. "Report of survey conducted at Mason & Hanger-Silas Mason Co., Inc." *Best Manufacturing Practices* (July). www.p2pays.org/ref%5C05/04934.pdf

Bureau of Labor Statistics (BLS), U.S. Department of Labor. 2010. *Nonfatal Occupational Injuries and Illnesses Requiring Days Away from Work for State Government and Local Government Workers, 2010* (retrieved October 4, 2011). www.bls.gov/news.release/pdf/osh2.pdf

Calex Electronics Limited. 2010. "Infrared Thermometry." *Understanding and Using the Infrared Thermometer* (retrieved June 23, 2011). www.calex.co.uk/downloads/application_guidance/understanding_and_using_ir.pdf

DoItYourself.com. www.doityourself.com/invt/1493865

Elliott, S. J. 2001. *Hardware for Active Control, Signal Processing for Active Control*. London: Academic Press.

Ernst, A., P., Koningsveld, and Henk F. Van Der Molen. 1997. "History and Future of Ergonomics in Building and Construction." *Ergonomics* (October 1) 40(10).

Helander, M. G. 1995. *A Guide to the Ergonomics of Manufacturing*. Bristol, PA: Taylor & Francis.

Humantech, Inc. 1996. *Ergonomic Design Guidelines for Engineers*. Ann Arbor, MI: Humantech, Inc..

International Labour Organization (ILO). 1986. ISO 5349-1986, *Mechanical Vibration – Guidelines for the Measurement and the Assessment of Human Exposure to Hand-Transmitted Vibration*. Geneva, Switzerland: ILO.

Jastrzebowski,W. 1857. "Rys historyczny czyli nauka o pracy, opartej na prawach poczerpni´tych z Nauki Przyrody" ("An Outline of Ergonomics or The Science of Work Based Upon the Truths Drawn from the Science of Nature") *Przyroda i Przemys* (*Nature and Industry*) pp. 29–32. Reprinted 1997. Warsaw: Central Institute for Labour Protection.

Karwowski, Waldemar, and William S. Marras. 1997. "Design and Management of Work Systems, Principles and Applications in Engineering," *Occupational Ergonomics*, p. 15.

Kroemer, Karl, Henrike Kroemer, and Katrin Kreomer-Elbert. 2002. *Ergonomics – How to Design for Ease and Efficiency*. 2d ed. Upper Saddle River, NJ: Prentice Hall International.

Lang, Susan S. 1999. "Vertical Split Keyboard Offers Lower Injury Risk for Typists." *Cornell News*. (retrieved August 2, 2005) www.news.cornell.edu/releases/Nov99/vertical.keyboard.ssl.html

Liberty Mutual Research Institute for Safety. 2009. *Workplace Safety Index*. Boston: Liberty Mutual Group.

———. 2009. *The Most Disabling Workplace Injuries Cost Industry an Estimated $52 Billion* (retrieved July 9, 2010). www.libertymutualgroup.com/omapps/ContentServer?c=cms_document&pagename=LMGResearchInstitute%2Fcms_document%2FShowDoc&cid=1138365240689

Liu, Kuang-Hung, Liang-Chieh Chen, Timothy Ma, Gowtham Bellala, and Kifung Chu. 2008. Report EECS 452, "Active Noise Cancellation Project" (retrieved June 20, 2011). www.personal.umich.edu/?gowtham/bellala_EECS452report.pdf

Lockheed Martin Company, Sandia Laboratories. 2010. *Human Factors Engineering* (retrieved August 11, 2010). www.reliability.sandia.gov/Human_Factor_Engineering/human_factor_engineering.html

LORD Corporation. 2011. *LORD MR Suspension Systems* (retrieved June 21, 2011). www.lord.com/Products-and-Solutions/Magneto-Rheological-%28MR%29.xml

M&R Companies. 2011. *Passport* (retrieved June 23, 2011). www.mrprint.com/en/ProductOverview.aspx?id=46

Melhorn, J. Mark. 1998. "Cumulative Trauma Disorders and Repetitive Strain Injuries—The Future." *Clinical Orthopedics & Related Research*. (June) 351:107–126.

Occupational Safety & Health Administration (OSHA). "Baggage Handling." *OSHA Ergonomics eTools*. www.osha.gov/SLTC/etools/baggagehandling/index.html

_____. 2002. OSHA 3080, *Hand and Power Tools* (retrieved October 4, 2011). setonresourcecenter.net/osha_pubs/osha3080.pdf

_____. 2003, revised 2009. *Ergonomics for the Prevention of Musculoskeletal Disorders—Guidelines for Nursing Homes*. www.osha.gov/ergonomics/guidelines/nursinghome/final_nh_guidelines.html/

_____. 2004. OSHA 3213-09N, *Ergonomics for the Prevention of Musculoskeletal Disorders, Guidelines for Poultry Processing* (retrieved July 9, 2010). www.osha.gov/SLTC/ergonomics/guidelines/poultryprocessing/poultryprocessing.html

_____. 2010. *Noise: Noise Induced Hearing Loss* (retrieved June 21, 2011). www.osha.gov/SLTC/teenworkers/hazards_noise_hearingloss.html

Omega Engineering Inc. 2011. *Industrial Non-Contact Infrared Thermometer/Transmitter* (retrieved June 22, 2011). www.omega.com/ppt/pptsc.asp?ref=os550a

Poynor, James. n.d. "Innovative Designs for Magneto-Rheological Dampers." *Overview of Magneto-Rheological (MR) Fluid Devices* (retrieved June 20, 2011). www.writing.engr.psu.edu/me5984/poynor.pdf

Pulat, B. Mustafa. 1997. *Fundamentals of Industrial Ergonomics*. Long Grove, IL: Waveland Press.

Putz-Anderson, Vern, ed. 1988. *Cumulative Trauma Disorders: A Manual for Musculoskeletal Disease of the Upper Limbs*. Bristol, PA: Taylor & Francis.

Radwin, R. G., T. J. Armstrong, D. B. Chaffin, G. D. Langolf, and J. W. Albers. 1990. "Hand-Arm Frequency-Weighted Vibration Effects on Tactility." *International Journal of Industrial Ergonomics* 6:75–82.

Radwin, R.G., and J.T. Haney. 1996. *An Ergonomics Guide to Hand Tools*. Fairfax, VA: AIHA Ergonomics Committee.

Rinaldi, Carlos, Tahir Cader, Thomas Franklin, and Markus Zahn. 2004. "Magnetic Nanoparticles in Fluid Suspension: Ferrofluid Applications." From James Schwartz and Cristian Contescu, eds., *Dekker Encyclopedia of Nanoscience and Nanotechnology*, Volume I, Chapter 123. New York: M. Dekker.

SawStop. 2011. *How It Works* (retrieved July 6, 2011). www.sawstop.com/products/industrial-cabinet-saw/

Scarfone, Karen, and John Padgette. 2008. Special Publication 800-121, *Guide to Bluetooth Security*. Recommendations of the National Institute of Standards and Technology (retrieved June 20, 2011). www.csrc.nist.gov/publications/nistpubs/800-121/SP800-121.pdf

Sitzman, Kathy, and Donald Bloswick. 2002. "Creative Use of Ergonomic Principles in Home Care—Clinical Concerns." *Home Healthcare Nurse*. (February) 20(2):98–103.

Smith, Sandy. 2004. "Off-the-Job Safety: Yardwork Leads to Blisters, Hand Injuries." *Occupational Hazards* (retrieved July 7, 2010). www.ehstoday.com/news/ehs_imp_37235/

Streeter, H. 1970. "Effects of Localized Vibration on the Human Tactile Sense." *American Industrial Hygiene Association Journal*. 31(1).

Terrel, R., and J. Purswell. 1976. "The Influence of Forearm and Wrist Orientation on Static Grip Strength as a Design Criterion for Hand Tools." *Proceedings of the Human Factors Society* 20:28–32.

Yassi, Annalee. 2000. "Work-Related Musculoskeletal Disorders, Nonarticular Rheumatism, Sports-Related Injuries, and Related Conditions." *Current Opinion in Rheumatology* (March) 12(2):124–130.

INDEX

A

Accidents. *See:* Incidents
ACGIH. *See:* American Conference of Governmental Industrial Hygienists (ACGIH)
Active hearing protectors, 143
Active noise control, 143
ADA. *See:* Americans with Disabilities Act (ADA), 1990
Aerobic processes, metabolic, 58–60
Affordances, 92–93
AIHA. *See:* American Industrial Hygiene Association (AIHA)
American Conference of Governmental Industrial Hygienists (ACGIH), 8
American Industrial Hygiene Association (AIHA), 2
American National Standards Institute (ANSI)
    ergonomics standards, 6–8
    Z10 Standard, 7
American Society of Safety Engineers (ASSE), ergonomics, 2, 7
Americans with Disabilities Act (ADA), 1990, 11
Anaerobic processes, metabolic, 58–60
ANS. *See:* Autonomic nervous system (ANS)
Anthropometry, 47–50, 130, 132, 145
Articulation Index, 87
ASC Z-365 Management of Work-Related Musculoskeletal Disorders (NSC), 7–8
ASSE. *See:* American Society of Safety Engineers (ASSE)
Audition. *See:* Hearing
Audits, 123
Autonomic nervous system (ANS), 54
Average, anthropometric design for, 48–49

B

Back
    anatomy, 51–52
    biomechanical modeling, 30–36
    injuries, 26–27
BCPE. *See:* Board of Certified Professional Ergonomists (BCPE)
Behavioral-based safety, 101
Benchmarking, ergonomic programs, 109–124
    across companies, 123–124
    metric classification, 111–112
    metric formation, 113–123
    OSHA record-keeping standard, 110–111
Best work zone, 135
Biomechanics, 29–36, 130, 145
Board of Certified Professional Ergonomists (BCPE), 22
BLS. *See:* Bureau of Labor Statistics (BLS)
Bluetooth technology, 144
Body
    dimensions, 47–50, 145
    systems, 50–58
    temperature, 56
Bone, 51
Box cutters, 143
British Standards Institute (BSI), 12
Bureau of Labor Statistics (BLS), 110, 127–128
Bursa injuries, 24
By-products, metabolic, 56

C

California, ergonomics standard, 9
Canada, ergonomics standards, 14–15
Canadian Standards Association (CSA), 15
Cardiovascular capacity, 60–62
Carpal tunnel syndrome, 25, 128–129, 139, 142, 145
Cartilage, 50–51
Cell phones, 144
Central nervous system (CNS), 53–54
Ceramic processing, 145
Circulatory system, 55–56
Closed-loop tasks, 98–99
CNS. *See:* Central nervous system (CNS)
Code of Federal Regulations, 29 CFR 1904, 110–111
Cognitive demands, 93–95
Cognitive tunnel vision, 95
Cold stress, 57, 137
Color coding, lighting and, 86–87
Common sense, 71–72
Complaint trending, 113–115, 118–120
Compliance, 2–4. *See also:* Inspection
Computer workstations, 7, 9, 50, 139–141
Confirmation, 95
Confusion matrix, 92
Connective tissue, 50–52

Control theory, 101
Controls, machine, 99–101
Courses. *See:* Training
CSA. *See:* Canadian Standards Association (CSA)
Culture, organizational, 113–115
Cumulative trauma disorders, 23–28. *See also:* Musculoskeletal disorders (MSDs)
Cutting tools, 142–143

D

De Quervain's Disease, 145
Decision making, 93–95
    auditory transmission criteria, 87–88
Design, safety. *See also:* Lockout/tagout (LOTO)
    ergonomic, 48–49, 65–66
    noise control, 136, 143
Diagnostic tools, 120–121
Disability, 121
The Diseases of Workers (Ramazzini), 22
Disks, spinal, 27, 35–36, 51–52
Documentation. *See:* Records
Dynamic biomechanical modeling, 32

E

Education. *See:* Training
EHRS. *See:* Ergonomic hazards and repetitive strains (EHRS)
Employees. *See also:* Training
    ergonomic hazards by industry, 138–145
    errors by, 94–95
    noise, 136, 143
    operator characteristics, 137
Energy requirements, physiologic, 57–58, 62–63
Equipment. *See also:* Personal protective equipment (PPE)
    hand tools, 131–137, 139–143
Ergonomic hazards and repetitive strains (EHRS)
    cumulative trauma disorders, 23–28
    defined, 127–128, 145–146
    by industry, 138–145
    musculoskeletal disorders, 112–113, 115, 127–129, 145
    risk factors, 23, 115, 127–129, 132–137, 145
Ergonomics, 21–108, 109–147. *See also:* Ergonomics regulations, Human factors engineering (HFE), and Work physiology
    benchmarking, 109–124
    best practices, 130–145
    hazards by industry, 138–145
    lifting equations, 36–40
    metrics, 115–123
    occupational biomechanics, 29–36
    operator characteristics, 137
    OSHA record-keeping standard, 110–111
    overview, 21–22, 127
    personal protective equipment, 133, 136, 138, 140
    posture, 48, 134–135, 138–139
    rapid upper limb assessment (RULA), 40–43
    sciences included in, 22–23
    tools for measuring body stress, 40
    workstation design, 7, 9, 50, 137–141
"Ergonomics Program Guidance Document" (AIHA), 7
Ergonomics Program Management Guidelines for Meatpacking Plants, 1990, 5
Ergonomics regulations, 1–15
    international standards, 12–15
    national guidelines, 2–9
    resources, 19
    state regulations, 9–10
    Workers' Compensation, 10–11
Errors, worker, 94–95
Evaluation. *See:* Inspection
Exertion, extreme, 53
Extension, 145
Extreme, anthropometric design for, 49
Eyestrain, 138, 140

F

Fatigue, 63–65
Fibrocartilage, 51, 53
Finger press, 145
Fitness testing, 58–62
Flexion, 145
Food processing industry, 140, 142, 145
Forklifts, 139
Fulwiler, R. D., 2
Functional capacity evaluation (FCE), 120–121

G

Gardening industry, 138–139
Garment industry, 139
Gilbreth, Frank & Lillian Moller, 22
Gloves
    ergonomics, 133
    gardening, 138
    poultry processing, 140
    vibration, 136
Grip strength, 34, 131–134

H

Hammers, 131, 134, 141
Hand activity level (HAL), 8
Hand tools, 131–136, 141–143
Handle sizing and types, 133–134
Hands, biomechanical modeling, 34
Hazards. *See also:* Ergonomic hazards and repetitive strains (EHRS)
    noise, 136, 143

Healthcare safety, 141–142
Hearing
    human factors engineering and, 87–91
    risks to, 136
Hearing programs. *See:* Noise control
Heart rate, 55, 60–61, 63
Heat stress, 57, 137
Herniated disks, 27
HFE. *See:* Human factors engineering (HFE)
HFES. *See:* Human Factors and Ergonomics Society (HFES)
HFES 100, *Human Factors Engineering of Computer Workstations* (ANSI/HFES), 7
HFES 200, *Software User Interface Standard* (ANSI/HFES), 7
HMTA. *See:* Human-machine task allocation (HMTA)
Human factors, 21
Human Factors and Ergonomics Society (HFES), 22
Human factors engineering (HFE), 69–103, 128–130
    behavioral-based safety, 101
    design issues, 74–77, 83, 95, 102–103
    documentation, 76
    hand tools, 131–136
    manual tracking, 98–101
    mental workloads, 95–97
    motor-performance demands, 97–98
    overview, 69–70, 73–74, 103, 128–130
    standards and performance, 102–103
    training issues, 101–102
    user performance mapping, 83–95
    workflow assessments, 77–83, 130
    workload assessment, 81–82
Human Factors Engineering of Computer Workstations (ANSI/HFES), 7
Human Factors Engineers (HFEs), 69
Human-machine interface, 131
Human–machine system transfer function, 101
Human-machine task allocation (HMTA), 131
Hyaline cartilage, 51

I

Identifiable sensations, 1
IEA. *See:* International Ergonomics Association (IEA)
ILO. *See:* International Labour Organization (ILO)
Imhotep, 21
In-line handles, 133–134
Incidents
    injury costs, 127
    injury reduction efforts, 118–120
Industrial hygiene, 145
Information transmittal, 91–92
Infrared (IR) noncontact thermometers, 145
Injury/complaint trending, 113–115, 118–120
Inspection, OSHA, 3
Intelligibility, speech, 87–91
International Labour Organization (ILO), 7, 12

International Organization for Standardization (ISO), ergonomics hazards, 12–13
Investigation. *See:* Inspection

J

Jackets, 132
Jastrzebowski, Wojciech, 127
Job analysis, 116–118
Job descriptions, 120

K

Keyboards, computer, 140–141
Kinesiology, 29, 145

L

Ladders, 138
Lagging metrics, 111–113
Leading metrics, 111–113, 116–121
Liberty Mutual Workplace Safety Index (WSI), 2009, 127–128
Lifting
    devices, 135, 139, 142
    guidelines, 138, 140
    metabolic costs, 58
    models, 34–36
    power zone, 49
    threshold limit values (TLV), 8
Ligaments, 24, 50–52
Lighting, as HFE design factor, 86–87, 136–137
Linkage analysis, 81
Linkage system models, 31–32
Long-term memory, 93
Lost work days, 110
Lung capacity, 54

M

Machine-human interface, 69–70, 131
Machine maintenance industry, 141
Magneto-rheological fluids (MRF), 144
Maine, video display terminal law, 9
Manual tracking, 98–101
Material handling, 138, 140, 145. *See also:* Ergonomics
Meatpacking industry, 5, 139
Medical safety, 141–142
Memory, 93
Mental fatigue, 64–65
Metabolic costs, 58
Metabolic system, 56–58
Metrics, 111–123
Michigan ergonomics standard, 9–10
Mobile phones. *See:* Cell phones
Modeling, biomechanical, 30–36

Motor control, 52–54
Motor-performance demands, 97–98
MRF. *See:* Magneto-rheological fluids (MRF)
Multiple-link coplanar static models, 31–32
Muscle
    fatigue, 53, 63–64
    injuries, 23
    strength modeling, 33
Musculoskeletal disorders (MSDs)
    defined, 127–128, 145
    metrics, 115–123

N

NACE. *See:* National Advisory Committee on Ergonomics (NACE)
NAICS. *See:* North American Industry Classification Systems (NAICS)
NAM. *See:* National Association of Manufacturers (NAM)
NASA. *See:* National Aeronautics and Space Administration (NASA)
National Academy of Social Insurance, 127
National Advisory Committee on Ergonomics (NACE), 4
National Aeronautics and Space Administration (NASA), 30
National Association of Manufacturers (NAM), 124
National Institute of Standards and Technology (NIST), 8
NCEP. *See:* Noise-canceling ear protection (NCEP)
Negative information bias, 95
Nerve disorders, 24
Neuromuscular system, 53–54
Neurovascular disorders, 24
Night work, 66–67
NIST. *See:* National Institute of Standards and Technology (NIST)
Noise-canceling ear protection (NCEP), 143
Noise control, ergonomic hazard, 87–89, 136, 143
North American Industry Classification Systems (NAICS), 110
The Nurse and Health Care Worker Protection Act (2009), 10

O

Occupational Health and Safety Management Systems (ANSI/AIHA), 7–8
Occupational Safety and Health Administration (OSHA)
    29 CFR 1904, 110–111
    ergonomic guidelines, 2–6
    General Duty Clause, 2–3
    hand tool guidelines, 131–132
    inspections, 3
    Voluntary Protection Program, 5–6
Office-based industries, 139–141. *See also:* Computer workstations
Open-loop tasks, 98
Operational analysis, 78–79
Organizational culture, 113–115
OSHA. *See:* Occupational Safety and Health Administration (OSHA)
Overexertion, 53
Oxygen uptake, 58–62

P

Packaging industry, 140
Performance mapping, 83–95
Peripheral nervous system, 53–54
Personal protective equipment (PPE), gloves, 133, 136, 138, 140
Personnel. *See:* Employees
Pharmaceutical industry, 139
Physical fitness testing, 58–62
Physiologic demands, 54–62
Pinch grip, 145
Pistol-grip handles, 133–134
Planar static biomechanical models, 30
Pliers, 143
Post-indication metrics. *See:* Lagging metrics
Posture, 134–135, 138
Poultry processing industry, 140
Power zone, 49
PPE. *See:* Personal protective equipment (PPE)
Pre-indication metrics. *See:* Leading metrics
Preemployment screening, 120–121
Preferred work zone, 135
"Prevention Through Design (PTD): Guidelines for Addressing Occupational Risks in Design and Redesign Processes" (ASSE 2010), 7

R

Radial deviation, 145
Ramazzini, Bernardino, 22
Range, anthropometric design for, 49
Raynaud's syndrome, 145
Reaction times, 97–98
Receiver operating characteristic curve, 90
Recordable injuries, 110, 113–115
Records
    ergonomics, 110–111
    human factors engineering, 76–77
Relative discrimination, 85–86
Repetitive motion, 53, 135
Resources, mental, 95–97
Respiratory system, 54

Risk assessment
    ergonomic hazards, 23, 115, 127–129, 132–137
    risk reduction, 145

S

Safety and Health Achievement Recognition Program (SHARP), 5–6
Safety and Health Management Guidelines, 1989, 5
Safety, behavioral-based, 101
Safety engineering. *See also: specific fields*
    human factors engineering, 70–73
    noise control, 136, 143
Seated workstations, 50, 137–138, 140–141
Self-report tools, 95
Sensations, identifiable, 84–85
Sensory memory, 93
Sensory thresholds, 84–86
Service and repair industry, 143
SHARP. *See:* Safety and Health Achievement Recognition Program (SHARP)
Shock absorbers and hydraulics, ergonomics of, 144
Short-term memory, 93
Shoulder flexion modeling, 33–34
SIC. *See:* Standard Industry Codes (SIC)
Signal detection, 89–91
Signal-to-noise ratio, 87–88
Sit-to-stand devices, 73
Site Specific Targeting (SST) Inspection Program (OSHA), 3
Skeletal muscular system, 52–53
Skeletal system, 50–52
Software User Interface Standard (ANSI/HFES), 7
Solis, Hilda, 2, 4
Sound levels. *See:* Noise control
Speech intelligibility, 87–91
Speech interference level (SIL), 88
Spinal column, 26–27, 51–52
Split keyboards, 140–141
SST. *See:* Site Specific Targeting (SST) Inspection Program
Stairs, 138
Standard Industry Codes (SIC), 110
Standing workstations, 138
Static strength modeling, 30–32
Stimulus intensity, 84–86
Stimulus signal detection, 89–91
Stress tests, 60–62
Symptom monitoring, 118–120

T

Task analysis, 79–81, 116–118
Task-human-environment-machine (THEM), 69–70, 74
Task linkages, 80–81
Task rotation, 121, 123
Taylor, Frederick Winslow, 22

Temperature
    infrared thermometers, 145
    metabolic, 56–57
    work area, 137
Tendonitis, 146
Tendons, 23–24, 50–51, 146
Tenosynovitis, 146
Testing. *See:* Inspection
Textile manufacturing industry, 142–143, 145
THEM. *See:* Task-human-environment-machine (THEM)
Thermal stress, 57
Thoracic outlet syndrome, 27–28
Threshold limit value (TLV), 8
Thresholds, sensory, 84–86
TLV. *See:* Threshold limit value (TLV)
Tools. *See also:* Equipment
    ergonomic designs, 121, 141–143
    hand, 131–136, 141–143
    maintenance, 134
    weight of, 135
Tracking tasks, 98–101
Training, ergonomics, 101–102, 121–122
Trend analysis, 113–115
Trigger finger, 25–26, 146
Triggers, 133–134

U

Ulnar deviation, 146
Uncertainty, 114
User expectations, 72
User performance mapping, 83–95

V

Vertical keyboards, 141
Vibration, 86, 132, 136
Video display terminals, 8, 140–141. *See also:* Computer workstations
Visual clarity, 86–87, 136–137
Visual display terminals, 8, 140–141
Voluntary Protection Program (VPP), 5–6

W

Warnings, failure of, 91–92
Washington, ergonomics standard, 9
White finger, 145
Woodworking industry, 144
Work
    classifications, 63, 65
    design considerations, 65–66
    physiologic demands, 54–56
    schedules and circadian rhythm, 66–67
    zones, 135

Work physiology, 47–67
    anthropometry, 58–60
    body systems, 50–58
    cardiovascular capacity, 60–62
    fatigue, 63–65
    overview, 47, 67
    schedules and circadian rhythm, 66–67
    work design, 65–66
Workers' Compensation
    costs, 127
    injury prevention and, 10–11
    and injury rate tracking, 115–116

Workload, 81–82, 95–97
Workplace assessment. *See:* Inspection
Workstation design, 7, 9, 50, 137–140
Wrists, biomechanical modeling, 34
WSI. *See:* Liberty Mutual Workplace Safety Index (WSI)

Z
Z10 Standard (ANSI), safety management, 7

CPSIA information can be obtained
at www.ICGtesting.com
Printed in the USA
FFOW04n0304110518
46577351-48586FF